# Enhancement of Grid-Connected Photovoltaic Systems Using Artificial Intelligence

Amal M. Abd El-Hameid • Adel A. Elbaset
Mohamed Ebeed • Montaser Abdelsattar

# Enhancement of Grid-Connected Photovoltaic Systems Using Artificial Intelligence

Springer

Amal M. Abd El-Hameid
Electrical Department
Faculty of Technology and Education
Sohag University
Sohag, Egypt

Adel A. Elbaset
Electrical Engineering
Department Faculty of Engineering
Minia University
El-Minia, Egypt

Mohamed Ebeed
Electrical Department
Faculty of Technology and Education
Sohag University
Sohag, Egypt

Montaser Abdelsattar
Electrical Engineering
South Valley University
Qena, Egypt

ISBN 978-3-031-29691-8 ISBN 978-3-031-29692-5 (eBook)
https://doi.org/10.1007/978-3-031-29692-5



This Springer imprint is published by the registered company Springer Nature Switzerland AG
The registered company address is: Gewerbestrasse 11, 6330 Cham, Switzerland

*Dedicated to our families and students*

# Preface

Nowadays, electricity consumption has been growing rapidly. As well as in the electric distribution networks, the R/X ratio is significantly higher than transmission systems; hence, power loss is high (about 10–13% of the generated power). Moreover, poor power quality, including the voltage profile and voltage stability issues, may arise.

Power system planners are becoming interested in enhancing the system performance by integrating Distributed Flexible AC Transmission Systems (D-FACTS) devices such as Distributed Static Compensator (D-STATCOM), Distribution Static Variable Compensator (D-SVC), and Unified Power Quality Conditioner (UPQC) along with Distributed Energy Resources (DERs), especially the renewable energy resources such as Photovoltaic (PV) and Wind Turbine (WT) generation units into the distribution system to withstand the electric grid issues. However, improper selection of the locations or the sizes of DERs and D-FACTS leads to greater losses and costs than without DERs or compensators. Therefore, the sites and sizes of the DERs and D-FACTS should be assigned optimally to enhance the system's performance significantly. The aim of this book is summarized as follows:

Assigning the optimal sites and sizes of the Photovoltaic Distributed Generation (PV-DG) and the D-STATCOM to enhance the system performance and reduce the system cost using developed optimization algorithms.

Using the recent optimization algorithms for solving the allocation problem such as the Marine Predators Algorithm (MPA), Equilibrium Optimizer (EO), Lightning Attachment Procedure Optimization (LAPO), Sine Cosine Algorithm (SCA), Ant Lion Optimizer (ALO), Whale Optimization Algorithm (WOA), and Slime Mould Algorithm (SMA), as well as improved versions algorithms, such as an Enhanced Sine Cosine Algorithm (ESCA) and a Modified Ant Lion Optimizer (MALO).

Validation of the proposed optimization algorithm methods using different standard IEEE test systems such as IEEE-33 bus, IEEE-69 bus, and IEEE-118 bus test systems, as well as real test systems such as the 30-bus of the East Delta Network (EDN), and the 94-bus practical distribution system situated in Portugal.

Determination of the locations and sizes of the PV-DG and the D-STATCOM optimally under deterministic and probabilistic conditions. In deterministic conditions, the hourly variations of the load demand and solar radiation are considered in four annual seasons, while in probabilistic conditions, three years of hourly historical data of solar irradiance and load demand are utilized to describe the uncertainties of the load demand and solar irradiance.

Investigating the techno-economic analysis of the optimal integration of the PV-DG and the D-STATCOM is carried out. The simulation results reveal that the optimal allocation of the PV-DGs and DSTATCOMs can enhance performance, reduce system power losses, improve voltage profiles, and considerably boost system stability. In addition, the results verify the superiority of the proposed optimization algorithms for solving the allocation problem of the PV-DGs and the D-STATCOMs compared with the traditional algorithms in terms of the considered objective function.

Sohag, Egypt — Amal M. Abd El-Hameid
El-Minia, Egypt — Adel A. Elbaset
Sohag, Egypt — Mohamed Ebeed
Qena, Egypt — Montaser Abdelsattar

# Contents


**1 Introduction** . . . 1
- 1.1 Distributed Generation . . . 1
- 1.2 Photovoltaic-Based Distributed Generation . . . 1
- 1.3 Distributed Static Compensator (D-STATCOM) . . . 2
- 1.4 Main Objectives of Book . . . 3
- 1.5 Book Outline . . . 3

**2 Literature Review and Power Quality Issues** . . . 5
- 2.1 Literature Survey . . . 5
- 2.2 Distributed Generation Background . . . 7
  - 2.2.1 Definition of Distributed Generation . . . 7
- 2.3 The Cost Structure of Distributed Generation Technologies . . . 9
- 2.4 Capital Costs . . . 10
- 2.5 Distributed Generation Applications and Technology . . . 10
  - 2.5.1 Distributed Generation Applications . . . 10
  - 2.5.2 Distributed Generation Technologies . . . 14
- 2.6 Power Quality Impact of PV-DG . . . 16
- 2.7 Distributed Static Compensator (D-STATCOM) . . . 16
- 2.8 Photovoltaic Energy Systems . . . 21
  - 2.8.1 Solar Photovoltaic . . . 22
  - 2.8.2 Grid-Connected PV System . . . 25
- 2.9 FACTS Devices . . . 26
  - 2.9.1 Types of FACTS Controllers . . . 26
  - 2.9.2 Shunt Compensation . . . 27
  - 2.9.3 Series Compensation . . . 27
  - 2.9.4 PV-DG Operation Modes . . . 27
- 2.10 D-FACTS Devices . . . 28
- 2.11 Integration of DGs with Distribution Networks . . . 29
- 2.12 Power Quality . . . 30
- 2.13 Power Quality Disturbances Classification . . . 30

2.14 Power Quality Issues . . . . . . . . 31
2.15 Classification of Power Quality Disturbances . . . . . . . . 32
2.15.1 Transient Power Quality Disturbance . . . . . . . . 32
2.15.2 Short-Duration Voltage Variations . . . . . . . . 33
2.15.3 Long-Duration Voltage Variations . . . . . . . . 34
2.15.4 Waveform Distortion . . . . . . . . 35
2.15.5 Flicker . . . . . . . . 36
2.15.6 Frequency Variations . . . . . . . . 36
2.16 Questions . . . . . . . . 36

**3 Stochastic Optimal Planning of Distribution System Considering Integrated Photovoltaic-Based DG and D-STATCOM** . . . . . . . . 39
3.1 Distributed Network . . . . . . . . 39
3.2 Backward/Forward Sweep (BFS) Algorithm . . . . . . . . 39
3.3 Forward/Backward Power Flow . . . . . . . . 43
3.4 Problem Formulation . . . . . . . . 44
3.4.1 The Objective Functions . . . . . . . . 44
3.5 System Constraints . . . . . . . . 51
3.5.1 Equality Constraints . . . . . . . . 51
3.5.2 Inequality Constraints . . . . . . . . 51
3.6 Uncertainty Modeling . . . . . . . . 53
3.6.1 Modeling of the Solar Irradiance . . . . . . . . 53
3.6.2 Load Demand Modeling . . . . . . . . 54
3.7 Sensitivity Analysis . . . . . . . . 54

**4 Optimal Allocation of Distributed Energy Resources Using Modern Optimization Techniques** . . . . . . . . 73
4.1 Metaheuristic Optimization Techniques . . . . . . . . 73
4.1.1 Marine Predators Algorithm (MPA) . . . . . . . . 73
4.1.2 Initialization . . . . . . . . 73
4.1.3 Assigning the Top Predator . . . . . . . . 74
4.1.4 The Brownian and Lévy Flight Orientations . . . . . . . . 74
4.1.5 FDAs Effect and Eddy Formation . . . . . . . . 76
4.1.6 Marine Memory . . . . . . . . 76
4.2 Equilibrium Optimizer (EO) . . . . . . . . 76
4.3 Lightning Attachment Procedure Optimization (LAPO) . . . . . . . . 80
4.4 Sine Cosine Algorithm (SCA) . . . . . . . . 84
4.4.1 Enhanced Sine Cosine Algorithm (ESCA) . . . . . . . . 84
4.5 Ant Lion Optimizer (ALO) . . . . . . . . 86
4.5.1 Random Movement of an Ant . . . . . . . . 86
4.5.2 Trapping in Antlion Pits . . . . . . . . 88
4.5.3 Sliding Ants Toward Antlions . . . . . . . . 89
4.5.4 Elitism . . . . . . . . 89
4.5.5 Catching Prey and Rebuilding the Pit . . . . . . . . 90
4.6 Modified Ant Lion Optimizer (MALO) . . . . . . . . 90

4.7 Whale Optimization Algorithm (WOA) . . . . . . . . . . . . . . . . . . . 91
4.7.1 Inspiration . . . . . . . . . . . . . . . . . . . . . . . . . . . . . . . . . 91
4.7.2 Mathematical Model and Optimization Algorithm . . . . . 92
4.8 Slime Mold Algorithm (SMA) . . . . . . . . . . . . . . . . . . . . . . . . 95
4.8.1 Approach Food . . . . . . . . . . . . . . . . . . . . . . . . . . . . . . 96
4.8.2 Wrap Food . . . . . . . . . . . . . . . . . . . . . . . . . . . . . . . . . 97
4.8.3 Grabble Food . . . . . . . . . . . . . . . . . . . . . . . . . . . . . . . 98

**5 Results and Discussion** . . . . . . . . . . . . . . . . . . . . . . . . . . . 101
5.1 IEEE-30 Bus Radial Distribution System . . . . . . . . . . . . . . . . . 101
5.1.1 Optimal Allocation of PV and D-STATCOM in RDS Using ALO Algorithm . . . . . . . . . . . . . . . . . . 101
5.2 IEEE-69 Bus and IEEE-118 Bus Radial Distribution System . . . . 104
5.3 Optimal Allocation of DER in RDS Using ALO and MALO . . . . 104
5.3.1 Case 1: Optimal Installation of PV System Under the Deterministic Condition . . . . . . . . . . . . . . . . 104
5.3.2 Case 2: Optimal Planning Under Uncertainties of System . . . . . . . . . . . . . . . . . . . . . . . . 107
5.4 Optimal Allocation of DER in RDS Using LAPO and EO Compared with WOA and SCA . . . . . . . . . . . . . . . . . . 116
5.4.1 IEEE-118 Bus Radial Distribution System . . . . . . . . . . . 116
5.5 Optimal Allocation of DER in RDS Using MPA and PSO . . . . . . 129
5.5.1 IEEE-94 Bus Radial Distribution System . . . . . . . . . . . . 129
5.6 Case 1: Allocation of the Hybrid System Without Considering the Uncertainty . . . . . . . . . . . . . . . . . . . . . . . . . . 131
5.6.1 Single Hybrid PV-DG and D-STATCOM . . . . . . . . . . . 132
5.6.2 Two Hybrid PV-DG and D-STATCOM Systems . . . . . . 132
5.7 Case 2: Allocation of the Hybrid System Considering the Uncertainty . . . . . . . . . . . . . . . . . . . . . . . . . . . . . . . . . . 134
5.7.1 Single Hybrid PV-DG and D-STATCOM Considering the Uncertainties . . . . . . . . . . . . . . . . . . . . 136
5.7.2 Two Hybrid PV-DG and D-STATCOM Considering the Uncertainties . . . . . . . . . . . . . . . . . . . . 138
5.8 Optimal Allocation of DER in RDS Using SMA . . . . . . . . . . . . . 139
5.8.1 IEEE-30 Bus Radial Distribution System . . . . . . . . . . . . 139
5.9 Optimal Allocation of DER in RDS Using ESCA and SCA . . . . . 142
5.9.1 IEEE-33 Bus Radial Distribution System . . . . . . . . . . . . 142
5.9.2 The First Case Study – 33-Bus System . . . . . . . . . . . . . 143
5.9.3 The Second Case Study – 69-Bus System . . . . . . . . . . . 144
5.9.4 Optimal Integral of PV-DG along with D-STATCOMS . . . . . . . . . . . . . . . . . . . . . . . . . . . 151
5.10 A Comparison Between the Proposed and Conventional Algorithms . . . . . . . . . . . . . . . . . . . . . . . . . 160
5.10.1 IEEE-118 Bus System . . . . . . . . . . . . . . . . . . . . . . . . . 160

**6 Conclusions and Future Work** . . . . . 163
6.1 Conclusions . . . . . 163

**Appendix** . . . . . 165
Appendix A . . . . . 165
Appendix B . . . . . 167
Appendix C . . . . . 168
Appendix D . . . . . 171
Appendix E . . . . . 174
Appendix F . . . . . 176
Appendix G . . . . . 177

**References** . . . . . 215

**Index** . . . . . 225

# Abbreviations

| | |
|---|---|
| ALO | Ant Lion Optimizer |
| CHP | Combined Heat and Power |
| CP | Custom Power |
| DEG | Distributed Energy Generation |
| DERs | Distributed Energy Resources |
| D-FACTS | Distributed Flexible AC Transmission Systems |
| DG | Distributed Generation |
| DSSC | Distributed Static Series Compensator |
| D-STATCOM | Distribution Static Compensator |
| D-SVC | Distribution Static Var Compensator |
| EO | Equilibrium Optimizer |
| ESCA | Enhanced Sine Cosine Algorithm |
| GA | Genetic Algorithm |
| GCP | Vector is Constructed by the Repetition |
| KCL | Kirchhoff Current Law |
| KVL | Kirchhoff Volt Law |
| LAPO | Lightning Attachment Procedure Optimization |
| LFD | Lévy Flight Distribution |
| LSF | Loss Sensitivity Factor |
| MALO | Modified Ant Lion Optimizer |
| MPA | Marine Predators Algorithm |
| O&M | Operation and Maintenance |
| PCC | Point of Common Coupling Connection |
| PDF | Probability Distribution Function |
| PQ | Power Quality |
| PV | Photovoltaic |
| RDERs | Renewable Distributed Energy Resources |
| RDS | Radial Distribution System |
| SCA | Sine Cosine Algorithm |
| SMA | Slime Mould Algorithm |

| | |
|---|---|
| STATCOM | Static Synchronous Compensator |
| TVD | Total Voltage Deviation |
| TVSI | Total Voltage Stability Index |
| UPQC | Unified Power Quality Conditioner |
| VD | Voltage Deviation |
| VLSI | Voltage Loss Sensitivity Index |
| VSC | Voltage Source Converter |
| VSI | Voltage Stability Index |
| WOA | Whale Optimization Algorithm |

# Symbols

| | |
|---|---|
| $H_1$, $H_2$, $H_3$ | Weighting factors |
| $P_{T,\ loss}$ | Total active power losses |
| $N_{PV}$ | Number of PV units |
| $N_I$ | Number of transmission lines |
| $N_C$ | Number of D-STATCOM |
| $V_{min}$ and $V_{max}$ | Lower and the upper voltage limits |
| $N_T$ | Number of branches |
| $P^2_{eff/q}$ | The total effective active power loads beyond |
| $Q^2_{eff/q}$ | The total effective reactive power loads beyond |
| $N_B$ | Total number of system buses |
| $G_{std}$ | Solar irradiance |
| X | Reactance of the feeder branch between buses n and r |
| R | Resistance of the feeder branch between buses n and r |
| $P_s$ | Active power transmitted through the branch between buses s and r |
| $Q_s$ | Reactive power transmitted through the branch between buses s and r |
| $V_s$ | The voltage of bus s |
| $V_r$ | The voltage of bus r |
| $Q_{Ds}$ | Reactive power injected by D-STATCOM |
| $P_{PV}$ | Active power inserted by PV unit |
| $n_p$ | The number of system branches |
| $P_{loss}(i)$ | The power loss of the branch *i* in the system |
| $C_E$ | Cost of energy loss in $/kWh |
| $C_{Grid}$ | Purchasing cost per electric power in $/kWh |
| $C_{DS}$ | Capital cost of D-STATCOM in $/kVAR |
| $C_{O\ \&\ M}$ | Operation and maintenance cost of the PV in $/kWh |
| CRF | Capital recovery factor |
| $\Gamma$ | Rate of interest on capital investment of the installed PV |

| | |
|---|---|
| $S$ | Lifetime of PV in years |
| $B$ | Asset rate of return |
| $N_D$ | Lifetime of d-statcom in years |
| $\text{Cost}_{\text{PV}}^{\text{Fixed}}$ | Fixed cost of the PV unit |
| $\text{Cost}_{\text{PV}}^{\text{Variable}}$ | Variable cost of the PV unit |
| $P_{\text{Grid}}$ | Imported power from the grid per hour |
| $P_{\text{PV}}$ | Generated power by the PV per hour |
| $P_{\text{sr}}$ | Rated power of the PV unit |
| $Q_{\text{DS}}$ | Rated reactive power of the d-statcom in kVAR |
| $P_{\text{loss}}$ | Active power losses |
| $P_{\text{Total \_ loss}}$ | Total power loss |
| $I_i^k$ | Current injection |
| $S_{\text{I}}$ | Specified power injection |
| $R$ | Reactance and resistance |
| $I_{\text{L}}^{(K)}$ | Current flow |
| $F_{\text{obj}}$ | Objective function |
| Nb−1 | Total number of system branches |
| $V_{\text{n}}$ | Voltage at bus (s) |
| $X_n$ and $R_n$ | Reactance and the resistance between bus (s) and bus (r) |
| $C_{\text{PV \& DST}}$ | Cost PV and d-statcom |
| $C_{\text{Loss}}$ | Annual energy loss cost |
| $K_{\text{loss}}$ | Cost of energy loss |
| NT | Number of the branches of the grid |
| $g_{\text{s}}$ | Solar irradiance in W/ m$^2$ |
| $G_{\text{std}}$ | Solar irradiance in a familiar natural environment of about 1000 W/m$^2$ |
| $X_{\text{c}}$ | Certain irradiance point |
| min $F(X, U)$ | Total annual cost |
| $P_{\text{Slack}}$ and $Q_{\text{Slack}}$ | Substation active and reactive powers |
| NPV and NDS | Number of PV and d-statcom units |
| $V_{\text{max}}$ | Maximum operating voltage |
| $V_{\text{min}}$ | Minimum operating voltage |
| $I_{\text{max}, n}$ | Maximum allowable current through branch (n) |
| P | Penetration level of the PV unit |
| $\mu_{\text{l}}$ and $\sigma_{\text{l}}$ | Mean and standard deviation of the load demand for each time |
| $\text{prob}_i^l$ | Probability occurrence of interval i |
| $l_i$ and $l_{i+1}$ | Starting and ending points of the interval i |
| Voc | Voltage open circuit |
| Isc | Short circuit current |
| IPh | Photo generated current [A] |
| Q | Electron charge $1.6 \times 10^{-19} c$ |
| K | Boltzmann's constant $1.38 \times 10^{-23} J/K$ |
| TC | Absolute temperature of the cell |
| Irs | Reverse saturation current [A] |
| Rs | Series resistance of the cell |
| Rsh | Shunt resistance of the cell |

# List of Figures


Fig. 2.1 Distributed electricity systems ........ 8
Fig. 2.2 Summary of DG applications [39] ........ 14
Fig. 2.3 Schematic diagram of D-STATCOM device ........ 18
Fig. 2.4 A Radial distribution system with D-STATCOM ........ 19
Fig. 2.5 Equivalent circuit of the PV cell ........ 22
Fig. 2.6 A flowchart to design a stand-alone system ........ 23
Fig. 2.7 A flowchart to design a grid-connected system ........ 25
Fig. 2.8 Examples of FACTS to shunt switcher ........ 27
Fig. 2.9 Examples of FACTS to series switcher ........ 28
Fig. 2.10 Power quality disturbances ........ 32
Fig. 2.11 Impulsive transient PQ disturbance ........ 33
Fig. 2.12 Oscillatory transient PQ disturbance ........ 33
Fig. 2.13 Short-duration voltage disturbances ........ 34
Fig. 2.14 Waveform distortion ........ 35
Fig. 2.15 Commutation notches ........ 35
Fig. 2.16 Voltage fluctuations ........ 36
Fig. 3.1 Flow chart of BFS load flow ........ 41
Fig. 3.2 A radial distribution network ........ 43
Fig. 3.3 The system with the PV and D-STATCOM ........ 44
Fig. 3.4 Single line diagram of the 10-bus radial distribution system ........ 57
Fig. 3.5 Single line diagram of the 12-bus radial distribution system ........ 63
Fig. 3.6 The convergence characteristic of the ALO objective function ........ 63
Fig. 3.7 Single line diagram of the 10-bus radial distribution system ........ 68
Fig. 4.1 Trajectories (**a**) Lévy flight motion and (**b**) Brownian motion ........ 74
Fig. 4.2 The flowchart of application the MPA for optimal PV-DG and D-STATCOM ........ 77

Fig. 4.3 Solution process of EO and sensitivity analysis for allocation of PV and D-STATCOM ............ 80
Fig. 4.4 Formation of the charges and leaders in the cloud ............ 81
Fig. 4.5 Solution process of LAPO and sensitivity analysis for allocation of PV and D-STATCOM ............ 83
Fig. 4.6 Inhabitance movement about the best solution based on cosine ............ 85
Fig. 4.7 The trend of the following status on the best placement counts on $y_1$ ............ 85
Fig. 4.8 Flowchart of ESCA ............ 87
Fig. 4.9 Flowchart of application of the MALO for optimal planning ............ 91
Fig. 4.10 Position vectors in 2D along with the expected positions ............ 93
Fig. 4.11 Spiral bubble-net behavior of humpback whales ............ 94
Fig. 4.12 Flowchart of WOA optimization technique ............ 96
Fig. 4.13 Evaluation of fitness ............ 98
Fig. 4.14 Flowchart of SMA optimization technique ............ 99
Fig. 5.1 Single line diagram of the (EDN) ............ 102
Fig. 5.2 Voltage profile of the system ............ 103
Fig. 5.3 Voltage Stability index of the system ............ 104
Fig. 5.4 Single line diagram of the IEEE-69 bus ............ 105
Fig. 5.5 Single line diagram of 118-bus system ............ 106
Fig. 5.6 The convergence characteristic of the ALO and MALO for power losses minimization with incorporating (**a**) single PV unit, (**b**) two PV units, and (**c**) three PV units ............ 110
Fig. 5.7 The seasonal hourly load profile ............ 111
Fig. 5.8 The seasons solar irradiance variations ............ 111
Fig. 5.9 The hourly output power of the PV unit ............ 113
Fig. 5.10 The output power of the first PV unit ............ 113
Fig. 5.11 The output power of the second PV unit ............ 114
Fig. 5.12 The power losses under uncertain conditions ............ 114
Fig. 5.13 The voltage profile of 69-bus system by incorporating the PV-DGs and D-STATCOMs in (**a**) spring, (**b**) summer, (**c**) autumn, and (**d**) winter ............ 115
Fig. 5.14 The voltage profile of 118-bus system by incorporating the PV-DGs and D-STATCOMs in (**a**) spring, (**b**) summer, (**c**) autumn, and (**d**) winter ............ 117
Fig. 5.15 The hourly output power of the single PV unit in 118-bus system ............ 117
Fig. 5.16 The output power of the first PV unit in the 118-bus system ............ 118
Fig. 5.17 The output power of the second PV unit in the 118-bus system ............ 118
Fig. 5.18 Single line diagram of 118-bus system ............ 118

Fig. 5.19 The VLSI for 118-bus system ........ 120
Fig. 5.20 The hourly load profile of four seasons ........ 120
Fig. 5.21 The hourly solar irradiance of four seasons ........ 120
Fig. 5.22 The power losses of four seasons at base case ........ 121
Fig. 5.23 The purchased power from grid at base case ........ 121
Fig. 5.24 The hourly output power of PV unit ........ 123
Fig. 5.25 The power losses of four seasons for case 1 ........ 124
Fig. 5.26 The purchased power for case 1 ........ 124
Fig. 5.27 The minimum total costs at different penetration levels ........ 126
Fig. 5.28 The output power of the first PV unit ........ 127
Fig. 5.29 The output power of the second PV unit ........ 127
Fig. 5.30 The power losses of four seasons for case 2 ........ 127
Fig. 5.31 The purchased power from grid for case 2 ........ 128
Fig. 5.32 The topology of the 94-bus system ........ 129
Fig. 5.33 The hourly load profiles ........ 130
Fig. 5.34 The hourly solar irradiance ........ 131
Fig. 5.35 The system power losses ........ 131
Fig. 5.36 The hourly PV-DG output power ........ 133
Fig. 5.37 The system losses for case 1 ........ 133
Fig. 5.38 The voltage profile of system for (**a**) spring, (**b**) winter, (**c**) summer, and (**d**) autumn ........ 135
Fig. 5.39 The power variations of the first PV-DG ........ 136
Fig. 5.40 The power variations of the second PV-DG ........ 136
Fig. 5.41 The load profile under uncertainty ........ 137
Fig. 5.42 The solar irradiance under uncertainty ........ 137
Fig. 5.43 The output power of PV unit under the uncertainty condition ........ 137
Fig. 5.44 Variation of the first PV output power under uncertainty condition ........ 138
Fig. 5.45 The output power of the second PV unit under uncertainty condition ........ 139
Fig. 5.46 The power losses under uncertainty condition ........ 139
Fig. 5.47 Single line diagram of the (EDN) ........ 140
Fig. 5.48 Voltage profile of all cases the system ........ 141
Fig. 5.49 The convergence characteristic of the SMA for power losses minimization ........ 141
Fig. 5.50 Single line diagram of IEEE-69 bus system ........ 142
Fig. 5.51 Single line diagram of IEEE-33 bus system ........ 142
Fig. 5.52 Voltage profile of the 33-bus system ........ 149
Fig. 5.53 Voltage stability index of the 33-bus system ........ 149
Fig. 5.54 Voltage profile of the 69-bus system ........ 150
Fig. 5.55 Voltage stability index of the 69-bus system ........ 151
Fig. 5.56 Change of total power loss single PV with iterations for the 33-bus system ........ 155

Fig. 5.57 Change of total power loss two PVs with iterations for the 33-bus system ... 155
Fig. 5.58 Change of total power loss three PVs with iterations for the 33-bus system ... 155
Fig. 5.59 Change of total of total power loss single PV with iterations for the 69-bus system ... 156
Fig. 5.60 Change of total power loss two PVs with iterations for the 69-bus system ... 156
Fig. 5.61 Change of total power loss three PVs with iterations for the 69-bus system ... 156
Fig. 5.62 Voltage profile with 33-bus system ... 159
Fig. 5.63 Voltage stability index with 33-bus system ... 159
Fig. 5.64 Voltage profile 69-bus system ... 159
Fig. 5.65 Voltage stability index with 69-bus system ... 159
Fig. 5.66 The convergence characteristics of the considered algorithms with integration single hybrid systems ... 161
Fig. 5.67 The convergence characteristics of the considered algorithms with integration two hybrid systems ... 161

# List of Tables


| | | |
|---|---|---|
| Table 5.1 | Parameters of system | 103 |
| Table 5.2 | Simulation results with or without PV units and D-STATCOM | 103 |
| Table 5.3 | The system specification of 69-bus systems and its initial power flow | 106 |
| Table 5.4 | The system specification and initial power flow | 107 |
| Table 5.5 | Results of optimal allocation of PV units for loss reduction | 107 |
| Table 5.6 | Comparative results for incorporating PV in the 69-bus system | 108 |
| Table 5.7 | The cost coefficients of the PV-DG and the D-STATCOM | 111 |
| Table 5.8 | The simulation results of inclusion of single PV-DG and D-STATCOM in 69-bus system considering uncertainties of system | 112 |
| Table 5.9 | The simulation results of inclusion of two PV-DGs and D-STATCOMs in 69-bus system considering uncertainties of system | 113 |
| Table 5.10 | The simulation results of inclusion of the PV-DGs and D-STATCOMs in 118-bus system considering uncertainties of system | 116 |
| Table 5.11 | The specification of the 118-bus system | 119 |
| Table 5.12 | The selected parameters of the optimization algorithms | 119 |
| Table 5.13 | A comparison of the investigated algorithms for case 1 under 20% penetration level | 122 |
| Table 5.14 | A comparison of the investigated algorithms for case 1 under 30% penetration level | 122 |
| Table 5.15 | A comparison of the investigated algorithms for case 1 under 40% penetration level | 123 |
| Table 5.16 | A comparison of the investigated algorithms for case 2 under 20% penetration level | 125 |

Table 5.17 A comparison of the investigated algorithms for case 2 under 30% penetration level ........................... 125
Table 5.18 A comparison of the investigated algorithms for case 2 under 40% penetration level ........................... 126
Table 5.19 Statistical results for case 1 with inclusion single PV and single D-STATCOM ........................................... 128
Table 5.20 Statistical results for case 2 with inclusion two PV units and two D-STATCOMs ............................................ 129
Table 5.21 the specification of the 94-bus system ............................. 130
Table 5.22 The simulation results of incorporating single hybrid PV-DG and D-STATCOM ........................................ 132
Table 5.23 The simulation results of incorporating two hybrid PV-DGs and D-STATCOMs ...................................... 133
Table 5.24 The obtained results with a single hybrid PV-DG and D-STATCOM ................................................ 136
Table 5.25 The obtained results with two hybrid PV-DGs and D-STATCOMs ............................................... 137
Table 5.26 The simulation results of the inclusion of the hybrid PV-DG and D-STATCOMs under the uncertainties of system ......................................................... 138
Table 5.27 Parameters of system ............................................. 140
Table 5.28 Simulation results with or without PV units and D-STATCOM ................................................ 140
Table 5.29 System specification and initial power flow ....................... 143
Table 5.30 Input parameters used in numerical simulations .................. 143
Table 5.31 The results of installing of PV units in the first system .......... 144
Table 5.32 Results of optimized allocation of PV units in 33 bus ........... 145
Table 5.33 Comparative results for incorporating single PV in 33-bus system .................................................. 146
Table 5.34 Comparative results for incorporating two PVs in 33-bus system .................................................. 147
Table 5.35 Comparative results for incorporating three PVs in 33-bus system .................................................. 148
Table 5.36 The results of installing of PV units in the second system ........................................................... 149
Table 5.37 Results of optimal allocation of PV-DGs ........................ 152
Table 5.38 Comparative results for incorporating single PV in 69-bus system .................................................. 153
Table 5.39 Comparative results for incorporating two PVs in 69-bus system .................................................. 154
Table 5.40 Comparative results for incorporating three PVs in 69-bus system .................................................. 155
Table 5.41 Simulation result with inclusion PV and D-STATCOM ......... 157
Table 5.42 Simulation result with inclusion PV and D-STATCOM ......... 158

Table 5.43 Comparative results for incorporating single hybrid PV and D-STATCOM in 118-bus system .......... 160
Table 5.44 Comparative results for incorporating two hybrid PV and D-STATCOM in 118-bus system .......... 161

# Chapter 1
# Introduction

## 1.1 Distributed Generation

Distributed generation (DG) refers to a variety of technologies that generate electricity at or near where it will be used, such as solar panels and combined heat and power. Distributed generation may serve a single structure, such as a home or business, or it may be part of a microgrid (a smaller grid that is also tied into the larger electricity delivery system), such as at a major industrial facility, a military base, or a large college campus. When connected to the electric utility lower-voltage distribution lines, distributed generation can help support the delivery of clean, reliable power to additional customers and reduce electricity losses along transmission and distribution lines [1].

## 1.2 Photovoltaic-Based Distributed Generation

Photovoltaic-based distributed generation (PV-DG) injected into the power system is considered a highly promising solution due to the advantage of clean energy use. However, the investigation of the optimal PV-DG allocation (site and size) is a significant task for power system requirements and assessment of PV potential. Recent research on PV-DG allocation is reviewed from two perspectives: (1) DG, optimization algorithms, and objectives; and (2) PV potential assessment methodologies [1 to 3].



A. M. A. El-Hameid et al., *Enhancement of Grid-Connected Photovoltaic Systems Using Artificial Intelligence*, https://doi.org/10.1007/978-3-031-29692-5_1

## 1.3 Distributed Static Compensator (D-STATCOM)

Distributed flexible AC transmission systems (D-FACTS) devices are static power-electronic devices installed in AC distribution networks to increase the power transfer capability, stability, and controllability of the networks through series and/or shunt compensation. D-FACTS comprises several devices, including D-STATCOM, Distribution Static Var Compensator (D-SVC), Distributed Static Series Compensator (DSSC) and Unified Power Quality Conditioner (UPQC) [4].

D-STATCOM is an efficient member of the D-FACTS devices that consists of a voltage source converter (VSC), a DC bus capacitor, a ripple filter, and a coupling transformer. VSC is constructed by using insulated gate bipolar transistors (IGBT) and metal-oxide-semiconductor field-effect transistors (MOSFET), where the switching of component is based on pulse-width modulation (PWM) sequences. D-STATCOM can inject or absorb both active and reactive power at a point of common coupling connection (PCC) by injecting a variable magnitude and phase angle voltage at PCC. D-STATCOM is incorporated in electric systems for enhancing the power quality, (a) improving the power factor, (b) balancing the loading, (c) mitigating the harmonic, (d) reactive power compensation, (e) reducing the power fluctuations of photovoltaic units minimizing the voltage sag, (f) mitigating the flicker in the electric system, and (g) minimizing the power losses [5].

Power systems are getting more advanced and complex due to diverse generation sources and transmission of power from these sources without modifying and adding additional transmission capability, which, in some cases, forces the system to operate under extremely overstressed situations. Additionally, it has become difficult to meet the requirement for reactive power and to maintain the bus voltage within adequate limits [6]. A power quality issue is an occurrence caused by an out-of-spec voltage, current, or frequency that causes a failure or misoperation in end-user equipment. Utility distribution networks, sensitive industrial loads, and critical commercial operations suffer from various types of outages and service interruptions, which can cost significant financial losses.

*The motivation of this book* is to enhance the performance of the system and reduce the total annual costs through optimal integration of the PV-DG and the D-STATCOM in distribution systems. The allocation problems of the PV-DGs and the D-STATCOMs are determined optimally by applying developed optimization algorithms. The PV-DGs and the D-STATCOMs are incorporated under deterministic and probabilistic conditions where the variations of the load demand and solar irradiance are considered.

## 1.4 Main Objectives of Book

*In this book*, the optimal allocation of the PV-DG and the D-STATCOM is determined using developed optimization algorithms. The main objectives of this book can be summarized as follows:

1. Applying developed optimization algorithms to assign the optimal locations and ratings of the hybrid PV-DG and the D-STATCOM to enhance the performance of the electrical systems and reduce the total annual cost.
2. Solving the allocation problem of the PV-DGs and the D-STATCOMs under a deterministic state.
3. Solving the allocation problem of the PV-DGs and the D-STATCOMs considering the uncertainties of the load demand and solar irradiance.
4. Verification of the proposed algorithms and testing on standard IEEE test systems such as IEEE-33 bus, IEEE-69 bus, and IEEE-118 bus test systems, as well as real test systems such as the 30-bus of the East Delta Network (EDN), and the 94-bus practical distribution system situated in Portugal.
5. Assessing and evaluating the optimal inclusion of the hybrid PV-DGs and the D-STATCOMs from a technical and economic perspective.

## 1.5 Book Outline

To achieve the above objectives, the present book is organized into six chapters, in addition to a list of references. The chapters are organized as follows:

*Chapter 1* This chapter presents a general overview, the background, and the motivation for the study.

*Chapter 2* This chapter presents a comprehensive survey of distributed energy resources (DERs), including their types, their benefits, and the applied methods for assigning their optimal locations and sizes, and also a comprehensive survey about the D-FACTS, including their types, their benefits, and the applied methods for assigning their optimal locations and sizes. This chapter also presents an overview of the D-STATCOM, including its topology, benefits, and the methods used for optimal integration of this device. An overview related to PV-DG is presented, including its principal function, benefits, and the applied optimization methods for including it optimally. Also, the power quality issues and the power quality disturbances classification are presented comprehensively in this chapter.

*Chapter 3* This chapter explains the backwards/forwards sweep (BFS) power flow algorithm and models of PV-DG and D-STATCOM in the BFS. Also, the considered objective function is discussed, including the total active power loss, the voltage deviations, the voltage stability index, and the total annual cost. The equality and inequality system constraints of the system are explained comprehensively.

*Chapter 4* This chapter presents the applied development optimization techniques, including MPA, ALO, EO, LAPO, SCA, ESCA, WOA, MALO, and SMA. Also, their inspiration, their mathematical modeling, and their step procedure are presented to determine the optimal sizes and sites of the PV-DGs and D-STATCOMs.

*Chapter 5* This chapter presents the simulation results and discussion for the optimal integration of the PV-DGs and D-STATCOMs in different distribution systems, including standard IEEE bus test systems such as IEEE-33 bus, IEEE-69 bus, IEEE-94 bus, and IEEE-118 bus test systems, and real test systems such as the 30-bus of the East Delta Network (EDN) and the 94-bus practical distribution system situated in Portugal.

*Chapter 6* This chapter presents the main conclusions of the book and the suggestions for future research topics.

# Chapter 2
# Literature Review and Power Quality Issues

## 2.1 Literature Survey

The main feature of the radial distribution network (RDN) is those power losses due to its high R/X ratio which leads to about 10–13% losses of the generated power [7]. Thus, two solutions can be applied is inclusion of distributed generators and/or compensators. Distributed generators (DGs) have recently been incorporated in RDN for electrical quality, economic anxiety, and environmental concerns. Distributed generators can be defined as electric power resources with limited capacity connected to electric systems closed to users. There are many types of DGs based on their technology including microturbines, fuel cells, combustion engines, wind turbines, geothermal, photovoltaic, and hydro systems [8, 9]. Photovoltaic-based distributed generation (PV-DG) is one of the new renewable energy resources that is widely incorporated in distribution networks for numerous benefits for providing environmental, technical, and economic benefits [10, 11].

Integration of renewable distributed generators and the shunt compensators is an effective solution from technical and economic perspectives. Several renewable-based technologies have been progressed to generate the required electricity, including wind turbine, biomass, solar thermal, solar PV, geothermal, and hydro systems [12]. Distribution flexible AC transmission systems (D-FACTS) are also widely embedded in the distribution grid [13]. D-FACTS include numerous controllers such as distribution static VAR compensator (D-SVC), unified power quality conditioner (UPQC), and distributed static compensator (D-STATCOM). The inclusion of DGs can yield several economic, technical, and environmental benefits. It can also reduce the power loss, reduce the voltage deviation, maximize the system reliability, decrease the emissions of greenhouse gases, and diminish the generation cost [10].



A. M. A. El-Hameid et al., *Enhancement of Grid-Connected Photovoltaic Systems Using Artificial Intelligence*, https://doi.org/10.1007/978-3-031-29692-5_2

The optimal integration of D-STATCOM in distribution grids had been presented in [14]; the immune algorithm had been employed to assign the siting and sizing of the D-STATCOM for cost and loss reduction. Harmony Search Algorithm had been implemented to optimize the site and size of the D-STATCOM for loss reduction. In [6], the bat algorithm was applied for optimizing the site and size of the D-STATCOM for losses reduction with multi-load levels. The binary gravitational search technique had been employed to assign the site and size of the D-STATCOM for enhancement [15]. They applied the imperialist competitive algorithm to determine the optimal location and size of the D-STATCOM for a multi-objective function under uncertainties of the load demand. A Bio-Inspired Cuckoo Search technique has been used for optimizing the allocation of the D-STATCOM for loss reduction under different load models. The differential evolution algorithm has been used to assign the location and size of D-STATCOM for cost and loss reduction [16–18]. The site and size of the D-STATCOM were allocated to reduce the losses and boost the stability and the voltage profile using a multi-objective sine cosine algorithm. The fuzzy-GA-based algorithm has been employed to optimize the allocation of the D-STATCOM for reducing losses and the total cost [19]. Numerous efforts have been presented to allocate the PV-DG in the power system. The particle swarm optimization (PSO) had been employed to allocate the D-STATCOM and DG for reducing the losses and supporting the voltage [20].

Incorporating renewable distributed energy resources (RDERs) in distribution networks such as solar PV units are challenging due to seasonal and daily variations of the solar irradiance and the weather variations that increase the uncertainty in the power system. Thus, it is mandatory to consider the uncertainties of the RDERs for efficient planning in the power system. Several efforts have been presented to integrate the DGs under the uncertainties. In [21], the optimal planning problem has been solved with RDERs considering uncertainties of solar radiation, load demand, and wind speed. The [22] assigned of DGs' optimal size and site under uncertainties of the RDER fuel price and future load growth. In Refs. [23, 24], the optimal reactive power dispatch problem has been solved under the uncertainties of renewable sources. In Refs. [25, 26], the optimal site and a rating of the RDERs have been assigned optimally under uncertainties of the solar radiation, wind speed, and load demand.

PV units convert the solar irradiance to DC power by its solar cells, and then an inverter is utilized to invert the DC voltage to an AC voltage. There are two options for connecting the PV system: the first installation is that the PV is connected to the electric grid which is known as an on-grid PV system connected, and the other connection is organized by feeding a stand-alone system known as an off-grid PV system. Several optimizations have been used for assigning the optimal capacity of PV units such as genetic algorithm, artificial bee colony [27], grasshopper optimization algorithm, moth flame optimizer, lightning attachment procedure optimization, backtracking search optimization [28], and ant lion optimization [29].

The reactive power compensation is an excellent solution for minimizing the system loss and enhancing the performance of the RDN. Reactive power compensation can be achieved by the installation of distributed flexible AC transmission system (D-FACTS) devices such as unified power quality conditioner (UPQC), distribution static VAR compensator (D-SVC), and D-STATCOM [15]. D-STATCOM is an effective controller where it is installed in a shunt to control the voltage magnitude of a certain bus by injecting or absorbing reactive power to the system through its voltage source convert connected to the system through coupling transformer [30]. D-STATCOM is installed for important technical and economic reasons related to the RDN such as power quality, loss minimization, harmonic mitigation, voltage stability, and voltage profile improvement [31].

Several excellent efforts have been done to assign the optimal placement and rating of the DSTATCM such as Taher and Afsari optimal location and setting of D-STATCOM in distribution networks used immune algorithm (IA) for reducing power congestion, energy loss, cost, and the voltage profile improvement [14], Devi and Geethanjali [32] applied particle swarm optimization (PSO) to assign the rating and location of the D-STATCOM and the distributed generation (DG) to alleviate the power loss and improvement the voltage profile. Ref. [6] applied the differential evolution algorithm (DE) to reduce the losses by the optimal reconfiguration and D-STATCOM. A binary gravitational search algorithm (BGSA) has been employed to solve the allocation problem of the D-STATCOM for reliability enhancement [15]. Cat swarm optimization has been used to assign the size and rating of the D-STATCOM to alleviate the power loss and improvement the voltage profile [33]. In Ref. [34], the optimal allocation of DG and D-STATCOM using a cuckoo searching algorithm was applied. In Ref. [35] a Particle Swarm Optimization algorithm for finding the optimal location and sizing of Distributed Generation and Distribution Static Compensator (D-STATCOM) with the aim of reducing the total power loss along with voltage profile improvement of Radial Distribution System is proposed. In Ref. [36], bacterial foraging optimization algorithm has been used to solve the allocation problem of DG and D-STATCOM for minimization of the loss and voltage deviation.

## 2.2 Distributed Generation Background

### *2.2.1 Definition of Distributed Generation*

Distributed generation refers to a variety of technologies that generate electricity at or near where it will be used, such as solar panels and combined heat and power. Distributed generation may serve a single structure, such as a home or business, or it may be part of a microgrid (a smaller grid that is also tied into the larger electricity delivery system), such as at a major industrial facility, a military base, or a large college campus [37]. When connected to the electric utility's lower-voltage distribution lines, distributed generation can help support the delivery of clean, reliable power to additional customers and reduce electricity losses along transmission and distribution lines [38, 39]. The concept of DG contrasts with the traditional centralized power generation concept, where the electricity is generated in large power stations and is transmitted to the end-users through transmission and distributions

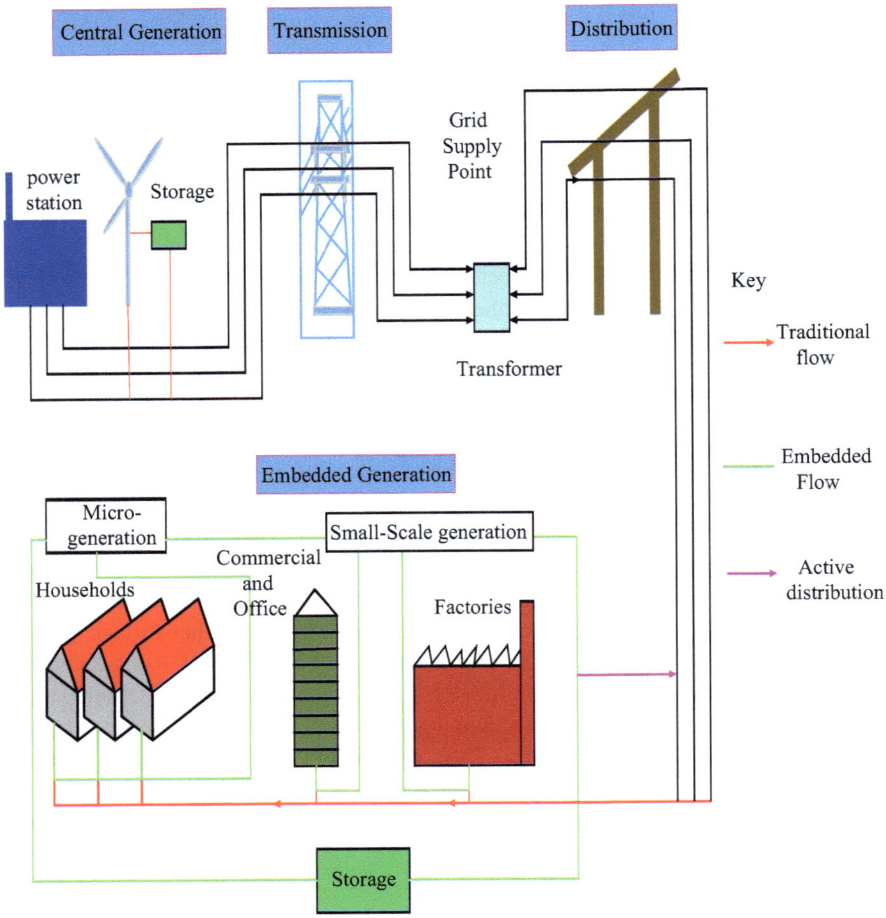


**Fig. 2.1** Distributed electricity systems

lines in Fig. 2.1. While central power systems remain critical to the global energy supply, their flexibility to adjust to changing energy needs is limited. Central power is composed of large capital-intensive plants and a transmission and distribution (T&D) grid to disperse electricity [40].

A distributed electricity system is one in which small and microgenerators are connected directly to factories, offices, and households and to lower voltage distribution networks. Electricity not demanded by the directly connected customers is fed into the active distribution network to meet demand elsewhere. Electricity storage systems may be utilized to store any excess generation. Large power stations and large-scale renewables, e.g. [41], offshore wind, remain connected to the high-voltage transmission network providing national backup and ensuring the quality of supply [38]. Again, storage may be utilized to accommodate the variable output of some forms of generation. Such a distributed electricity system is represented in Fig. 2.1 below.

In the residential sector, common distributed generation systems include:

- Solar photovoltaic panels
- Small wind turbines
- Natural-gas-fired fuel cells
- Emergency backup generators, usually fuelled by gasoline or diesel fuel

In the commercial and industrial sectors, distributed generation can include resources such as:

- Combined heat and power systems
- Solar photovoltaic panels
- Wind
- Hydropower
- Biomass combustion or cofiring
- Municipal solid waste incineration
- Fuel cells fired by natural gas or biomass

Reciprocating combustion engines, including backup generators, may be fuelled by oil [42].

## 2.3 The Cost Structure of Distributed Generation Technologies

The direct costs of distributed generation to customers include the installed cost of the equipment, fuel costs, nonfuel operation, and maintenance (O&M) expenses, and certain costs that the customers' utility imposes [43]. To make this comparison of costs most useful, the following cost data assume that for each technology, there are used an installed capacity, a rate of utilization, and (in some cases) a geographic location that would be suitable for serving the electricity needs of individual customers. For example, the costs for the wind turbine discussed here are for a size that might be used in a small rural business (such as a farm) in a location with favorable wind resources [44].

On that basis, data compiled from various industry and government sources describe the current costs of the most common types of electricity generation technologies. Data for a combined-cycle unit are presented as well; as the largest source of additional electricity from utilities and independent power producers, combined-cycle systems provide a representative benchmark against which the costs of other technologies can be measured [43].

## 2.4 Capital Costs

The costs of acquiring and installing generating units vary widely, depending on technology, capacity, and other factors. The US Department of Energy estimates that the typical installed capital costs for distributed generators range from under $1000 per kilowatt for a combustion turbine to almost $7000 per kilowatt for a solar photovoltaic system. Among small-capacity technologies, internal combustion engines (fuelled by diesel and gasoline) have the lowest capital costs and highest operating costs. Renewable technologies (using wind and solar power) have the highest capital costs and lowest operating costs [44].

## 2.5 Distributed Generation Applications and Technology

### *2.5.1 Distributed Generation Applications*

Distributed generation (DG) is currently being used by some customers to provide some or all their electricity needs. There are many different potential applications for DG technologies. For example, some customers use DG to reduce demand charges imposed by their electric utility, while others use it to provide primary power or reduce environmental emissions. DG can also be used by electric utilities to enhance their distribution systems. Many other applications for DG solutions exist. The following is a list of those of potential interest to electric utilities and their customers [43, 45].

#### 2.5.1.1 Continuous Power

In this application, the DG technology is operated at least 6000 hours a year to allow a facility to generate some or all its power on a relatively continuous basis. Important DG characteristics for continuous power include:

- High electric efficiency
- Low variable maintenance costs
- Low emissions

Currently, DG is being utilized most often in a continuous power capacity for industrial applications such as food manufacturing, plastics, rubber, metals, and chemical production. Commercial sector usage, while a fraction of total industrial usage, includes sectors such as grocery stores and hospitals [39].

#### 2.5.1.2 Combined Heat and Power (CHP)

Also referred to as cooling, heating, and power or cogeneration, this DG technology is operated at least 6000 hours per year to allow a facility to generate some or all its power. A portion of the DG waste heat is used for water heating, space heating, steam generation, or other thermal needs. In some instances, this thermal energy can also be used to operate special cooling equipment. Important DG characteristics for combined heat and power include [46]:

- High useable thermal output (leading to high overall efficiency)
- Low variable maintenance costs
- Low emissions

CHP characteristics are like those of continuous power, and thus the two applications have almost identical customer profiles, though the high thermal demand here is not necessary for continuous power applications. As with continuous power, CHP is most used by industry clients, with a small portion of overall installations in the commercial sector [39].

#### 2.5.1.3 Peaking Power

In a peaking power application, DG is operated between (200–3000) hours per year to reduce overall electricity costs. Units can be operated to reduce the utility demand charges, to defer buying electricity during high-price periods, or to allow for lower rates from power providers by smoothing site demand. Important DG characteristics for peaking power include [46]:

- Low installed cost
- Quick start-up
- Low fixed maintenance costs

Peaking power applications can be offered by energy companies to clients who want to reduce the cost of buying electricity during high-price periods. Currently, DG peaking units are being used mostly in the commercial sector, as load profiles in the industrial sector are relatively flat. The most common applications are in educational facilities, lodging, miscellaneous retail sites, and some industrial facilities with peaky load profiles [39].

#### 2.5.1.4 Green Power

DG units can be operated by a facility to reduce environmental emissions from generating its power supply. Important DG characteristics for green power applications include:

- Low emissions
- High efficiency
- Low variable maintenance costs

Green power could also be used by energy companies to supply customers who want to purchase power generated with low emissions [39, 47].

#### 2.5.1.5 Premium Power

DG is used to provide electricity service at a higher level of reliability and/or power quality than typically available from the grid. The growing premium power market presents utilities with an opportunity to provide a value-added service to their clients. Customers typically demand uninterrupted power for a variety of applications, and for this reason, premium power is broken down into three further categories [48].

#### 2.5.1.6 Emergency Power System

This is an independent system that automatically provides electricity within a specified time frame to replace the normal source if it fails. The system is used to power critical devices whose failure would result in property damage and/or threatened health and safety. Customers include apartment, office and commercial buildings, hotels, schools, and a wide range of public gathering places [39].

#### 2.5.1.7 Standby Power System

This independent system provides electricity to replace the normal source if it fails and thus allows the customer entire facility to continue to operate satisfactorily. Such a system is critical for clients like airports, fire and police stations, military bases, prisons, water supply and sewage treatment plants, natural gas transmission and distribution systems, and dairy farms [49].

#### 2.5.1.8 True Premium Power System

Clients who demand uninterrupted power, free of all power quality problems such as frequency variations, voltage transients, dips, and surges, use this system. Power of this quality is not available directly from the grid – it requires both auxiliary power conditioning equipment and either emergency or standby power. Alternatively, a DG technology can be used as the primary power source, and the grid can be used as a backup [50]. This technology is used by mission critical systems like airlines, banks, insurance companies, communications stations, hospitals, and nursing homes.

Important DG characteristics for premium power (emergency and standby) include:

- Quick start-up
- Low installed cost
- Low fixed maintenance costs

#### 2.5.1.9 Transmission and Distribution Deferral

In some cases, placing DG units in strategic locations can help delay the purchase of new transmission or distribution systems and equipment such as distribution lines and substations. A detailed analysis of the life-cycle costs of the various alternatives is critical, and issues relating to equipment deferrals must also be examined closely. Important DG characteristics for transmission and distribution deferral (when used as a "peak deferral") include [39]:

- Low installed cost
- Low fixed maintenance costs

#### 2.5.1.10 Ancillary Service Power

DG is used by an electric utility to provide ancillary services (interconnected operations necessary to affect the transfer of electricity between the purchaser and the seller) at the transmission or distribution level. In markets where the electric industry has been deregulated and ancillary services unbundled (in the United Kingdom, for example), DG applications offer advantages over currently employed technologies [39].

Ancillary services include spinning reserves (unloaded generation, which is synchronized and ready to serve additional demand) and non-spinning, or supplemental, reserves (operating reserve is not connected to the system but can serve demand within a specific time or interruptible demand that can be removed from the system within a specified time). Other potential services range from transmission market reactive supply and voltage control, which uses generating facilities to maintain a proper transmission line voltage, to distribution level local area security, which provides back up power to end-users in the case of a system fault [48].

The characteristics that may influence the adoption of DG technologies for ancillary service applications will vary according to the service performed and the ultimate shape of the ancillary service market [51, 52]. Figure 2.2 summarizes the different kinds of DG applications.

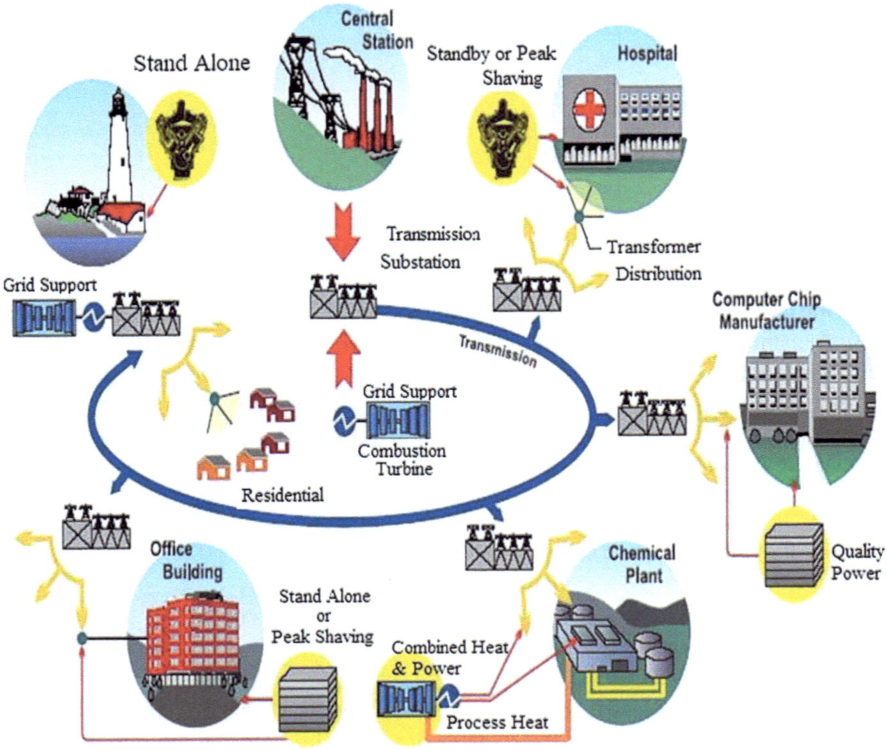


**Fig. 2.2** Summary of DG applications [39]

## 2.5.2 *Distributed Generation Technologies*

### 2.5.2.1 Reciprocating Engines

The distributed generation technology was developed more than a century ago and is still widely utilized in a broad array of applications. The engines range in size from less than 5 to over 5000 kW and use diesel, natural gas, or waste gas as their fuel source. Development efforts remain focused on improving efficiency and reducing emission levels. Reciprocating engines are being used primarily for backup power, peaking power, and in cogeneration applications [48].

### 2.5.2.2 Microturbines

A new and emerging technology, microturbines are currently only available from a few manufacturers. Other manufacturers are looking to enter this emerging market, with models ranging from 30 to 200 kW. Microturbines promise low emission levels, but the units are currently relatively expensive. Obtaining reasonable costs and demonstrating reliability will be major hurdles for manufacturers. Microturbines are just entering the marketplace, and most installations are for the purpose of testing the technology [39].

#### 2.5.2.3 Industrial Combustion Turbines

A mature technology, combustion turbines range from 1 MW to over 5 MW. They have low capital cost, low emission levels, but also usually low electric efficiency ratings. Development efforts are focused on increasing efficiency levels for this widely available technology. Industrial combustion turbines were being used primarily for peaking power and in cogeneration applications [39].

#### 2.5.2.4 Photovoltaic

Commonly known as solar panels, (PV) panels are widely available for both commercial and domestic use. Panels range from less than 5 kW, and units can be combined to form a system of any size. They produce no emissions and require minimal maintenance. However, they can be quite costly. Less expensive components and advancements in the manufacturing process are required to eliminate the economic barriers now impeding wide-spread use of PV systems. PV is currently being used primarily in remote locations without grid connections and to generate green power [49].

#### 2.5.2.5 Fuel Cells

Fuel cells not only are very efficient but also have very low emission levels. A fuel cell operates like a battery. It supplies electricity by combining hydrogen and oxygen electrochemically without combustion. However, while the battery is a storage device for energy that is eventually used up and/or must be recharged, the fuel cell is permanently fed with fuel and an oxidant, so that the electrical power generation continues. The final product is pure water; the electrochemical reaction generates electricity and heat without a flame ("cold combustion"). A single cell provides less than one volt, so a series of fuel cells are normally "stacked" one on another to increase the power output. The basic fuel cell has two electrodes separated by an electrolyte [39].

One of the electrodes (the anode) is supplied with the fuel (for example, hydrogen or natural gas). The second electrode (the cathode) is supplied with oxygen by simply pumping air in. The few fuel cells currently being used provide premium power. There are several types of fuel cells. Proton exchange membrane fuel cells are now days the most commercially available type. They have the highest energy density per volume rate, and their prices are expected to fall fast because they are being adapted by the automotive industry for transportation use [53].

#### 2.5.2.6 Wind Turbine Systems

Wind turbines are currently available from many manufacturers and range in size from less than 5 to over 1000 kW. They provide a relatively inexpensive (compared to other renewables) way to produce electricity, but as they rely upon, the variable and somewhat unpredictable wind are unsuitable for continuous power needs. Development efforts look to pair wind turbines with battery storage systems that can provide power in those times when the turbine is not turning. Wind turbines are being used primarily in remote locations not connected to the grid and by energy companies to provide green power [39].

## 2.6 Power Quality Impact of PV-DG

The integration of PV-DG in power systems can alleviate overloading in transmission lines, provide peak shaving, and support the general grid requirement. However, improper coordination, location, and installation of PV-DG may affect the power quality of power systems. Most conventional power systems are designed and operated such that generating stations are far from the load canters and use the transmission and distribution system as pathways. The normal operation of a typical power system does not include generation in the distribution network or in the customer side of the system [54].

However, the integration of PV-DG in distribution systems changes the normal operation of power systems and poses several problems which include possible bidirectional power flow, voltage variation, breaker no coordination, alteration in the short circuit levels, and islanding operation [55, 56]. Therefore, studies are required to address the technical challenges caused by DG integration in distribution systems. The interconnection device between the DG and the grid must be planned and coordinated before connecting any DG.

## 2.7 Distributed Static Compensator (D-STATCOM)

With the growth and development of power grids, optimal utilization of electric networks is very important. Because of the high cost of construction and development of power networks, mitigation of existing issues, such as excessive power losses, voltage profile problems, voltage instabilities, reliability problems, etc., is inevitable. To obviate these problems, distribution synchronous static compensator (D-STACTOM) as a shunt compensator device can be used in electric distribution networks. The optimal location and size of D-STACOM should be determined because of economic viability, required quality, reliability, and availability [57].

In recent years, several theses have concentrated on the techniques used for finding optimal location and size of the D-STATCOM units considering different aspects. However, to date, no review chapter has been published in this field. The chapter presents an up-to-date survey of the literature on the optimal allocation of D-STATCOM in distribution networks. The existing research works have been classified into five categories including analytical methods, artificial neural network-based approaches, metaheuristic methods, sensitivity approaches, and a combination of sensitivity approaches and metaheuristic methods. Moreover, it was found that in the D-STATCOM allocation problem, the objectives may be the alleviation of power loss, mitigation of voltage deviations, improvement of reliability metrics, and enhancement of voltage stability [58].

Increased loading and the need for economic efficiency of electric power networks have prompted companies to use transmission and distribution networks at the highest possible efficiency and loading [59]. Some challenges associated with these networks include stability issues, high losses, and excessive voltage drops in buses [60 - 62]. These challenges usually occur when there is an indiscriminate increase in nonlinear loads [63]. A voltage drop in the distribution network may be caused by distribution lines with high impedance or growing loads in three-phase or unbalanced loads. Today's advanced distribution networks take on a more sophisticated form due to the incorporation of distributed generation units that redirect the flow of current through the lines [64]. Nowadays, there is a general consensus on this idea that power electronics devices and methods are more suitable than traditional methods that were working based on electromechanical technologies and have low speed and high cost [65]. Using flexible alternating current transmission system (FACTS), devices is inevitable for optimal utilization of current electric networks [66, 67].

Power electronics-based controllers in power distribution systems help to provide energy with an appropriate quality for subscribers [68, 69]. In general, custom power (CP) devices, which are similar to FACTS devices, are a useful solution to address the problem of interruptions and poor power quality in power networks [70] . Although FACTS and CP devices share a common technical base, they have different performance goals. FACTS devices are used in the transmission, while CP devices are used in distribution. CP devices are especially effective in system reliability and power quality. Distribution static compensator (D-STATCOM) is a shunt CP device, used in distribution networks. It is used to improve power quality (i.e., power factor, voltage profile, voltage stability). The ability to inject and absorb reactive power with very fast, dynamical responses made its application very wide within the field. D-STATCOM injects current into the system at the point of common coupling, which helps in power factor correction, harmonic filtering, etc. [31]. A review of the literature indicates that potential applications of D-STATCOM include reactive power compensation, voltage support, circulating excess power among the phases, and reduction of fluctuations caused by photovoltaic systems [71].

Fig. 2.3 Schematic diagram of D-STATCOM device

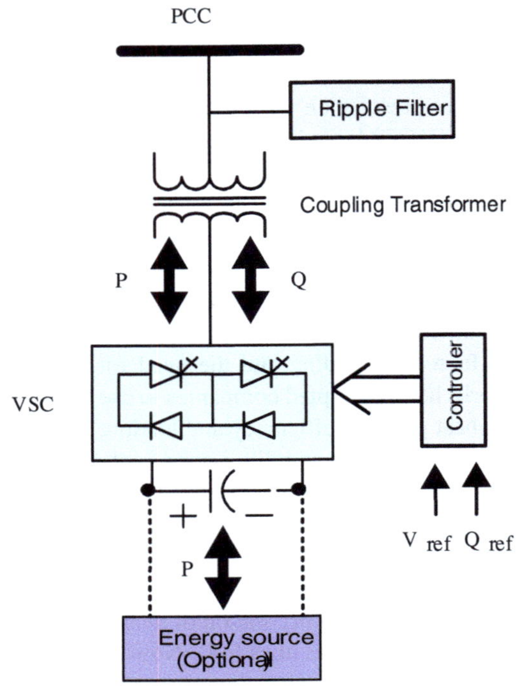


New members of FACTS controllers have been emerged due to continuous progress of power electronic devices. D-STATCOM is a developed controller based on voltage source converter (VSC). D-STATCOM can inject or absorb both active and reactive power at a point of common coupling connection (PCC) by injecting a variable magnitude and phase angle voltage at PCC. D-STATCOM is incorporated in electric systems for enhancing the power quality, improving the power factor, balancing the loading, mitigating the harmonic, reactive power compensation, reducing the power fluctuations of photovoltaic units minimizing the voltage sag, mitigating the flicker in the electric system, and minimizing the power losses [31, 71–74].

D-STATCOM consists of voltage source converter, DC bus capacitor, ripple filter, and coupling transformer as shown in Fig. 2.3. VSC is constructed by using insulated gate bipolar transistors (IGBT), Gate Turn-off thyristor (GTO), and MOSFET where the switching of component is based on pulse-width modulation (PWM) sequences. The coupling transformer is utilized for matching the inverter voltage with the bus voltage. The D-STATCOM topologies are categorized based on three-phase three-wire (3P3W) and three-phase four-wire (3P4W) as illustrated in [75, 76].

D-STATCOM can exchange active and reactive current with the network. A steady-state modeling D-STATCOM has presented [75]. Figure. 2.4 shows D-STATCOM controller which included in radial distribution system at bus $n + 1$ where D-STATCOM inject or absorb $I_D$ at this bus. By applying KVL, the voltage at bus $n + 1$ can be obtained as:

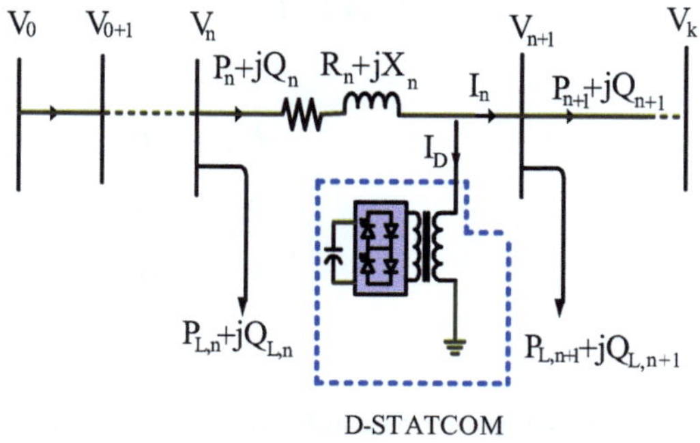


**Fig. 2.4** A Radial distribution system with D-STATCOM

$$V_{n+1} \angle\theta_{n+1} = V_n \angle\theta_n - (R_n + jX_n)\left(I_n \angle\delta + I_D \angle\left(\theta_{n+1} \pm \frac{\pi}{2}\right)\right) \tag{2.1}$$

where

$V_{n+1} \angle \theta_{n+1}$ is voltage of bus n + 1 after inclusion D-STATCOM,

$I_D$ is the injected current by D-STATCOM, and

$I_n$ is the line current after inclusion of DSTACOM.

Equation (2.1) represents the essential idea for modeling D-STATCOM which can be solved by separating it to real and imaginary terms as follows:

$$\begin{aligned} V_{n+1}\text{Cos}(\theta_{n+1}) &= Re(V_n \angle\theta_n) - Re(I_n \angle\delta(R_n + jX_n)) \\ &\quad + X_n I_D \text{Sin}\left(\theta_{n+1} + \frac{\pi}{2}\right) - R_n I_D \text{Cos}\left(\theta_{n+1} + \frac{\pi}{2}\right) \end{aligned} \tag{2.2}$$

where $R_n + jX_n$ is the reactance and resistance between bus $(_{n+1})$, $I_D$ is the injected current by D-STATCOM, $I_n$ is the line current after inclusion of DSTACOM, *Re* is the real part, and $\theta_{n+1}$ is the vase defiance between voltage of bus $V_n$, $V_{n+1}$

$$\begin{aligned} V_{n+1}\text{Sin}(\theta_{n+1}) &= Im(V_n \angle\theta_n) - Im(I_n \angle\delta(R_n + jX_n)) \\ &\quad - X_n I_D \text{Cos}\left(\theta_{n+1} + \frac{\pi}{2}\right) - R_n I_D \text{Sin}\left(\theta_{n+1} + \frac{\pi}{2}\right) \end{aligned} \tag{2.3}$$

Equations (2.2) and (2.3) can be simplified as follows:

$$a\text{Cos}x_2 = k_1 - b_1 x_1 \text{Sin}x_2 - b_2 x_1 \text{Cos}x_2 \tag{2.4}$$

$$a\text{Sin}x_2 = k_2 - \text{b}_2 x_1 \text{Sin}x_2 + \text{b}_1 x_1 \text{Cos}x_2 \tag{2.5}$$

where

$$k_1 = Re(V_n \angle\theta_n) - Re(I_n \angle\delta(R_n + jX_n)) \tag{2.6}$$

$$k_2 = Im(V_n \angle\theta_n) - Im(I_n \angle\delta(R_n + jX_n)) \tag{2.7}$$

$$a = V_{n+1} \tag{2.8}$$

$$b_1 = -R_n$$

$$b_2 = -X_n$$

$$x_1 = I_{\mathrm{D}}$$

$$x_2 = \theta_{n+1}$$

Equations (2.4) and (2.5) can be rewritten as follows:

$$x_1 = \frac{a\mathrm{Cos}x_2 - k_1}{-b_1\mathrm{Sin}x_2 - b_2\mathrm{Cos}x_2} \tag{2.9}$$

$$x_1 = \frac{a\mathrm{Sin}x_2 - k_2}{-b_2\mathrm{Sin}x_2 + b_1\mathrm{Cos}x_2} \tag{2.10}$$

By equating Eqs. (2.9) and (2.10)

$$(k_1b_2 - k_2b_1)\mathrm{Sin}(x_2) + (-k_1b_1 - k_2b_2)\mathrm{Cos}(x_2) + ab_1 = 0 \tag{2.11}$$

The previous equation can be simplified as:

$$(d_1^2 + d_2^2)x^2 + (2d_1ab_1)x + (a^2b_1^2 - d_2^2) = 0 \tag{2.12}$$

where it is assumed that:

$$x = \mathrm{Sin}(x_2)$$

$$d_1 = (k_1b_2 - k_2b_1)$$

$$d_2 = (-k_1b_1 - k_2b_2)$$

Hence, Eq. (2.12) can be solved as follows:

$$x = \frac{-B \pm \sqrt{B^2 - 4AC}}{2A} \tag{2.13}$$

where

$$A = (d_1^2 + d_2^2)$$

$$B = (2d_1ab_1)$$

$$C = \left(a^2 b_1^2 - d_2^2\right)$$

Hence,

$$\theta_{n+1} = \mathrm{Sin}^{-1}(x) \tag{2.14}$$

The value of $I_\mathrm{D}$ can be obtained simply from (2.6) or (2.7). The voltage at PCC, the D-STATCOM current, and injected reactive power by D-STATCOM can be found as follows:

$$\overrightarrow{V_{n+1}} = V_{n+1} \angle \theta_{n+1} \tag{2.15}$$

$$\overrightarrow{I_{\mathrm{DStatcom}}} = I_{\mathrm{DStatcom}} \angle \left(\theta_{k+1} + \frac{\pi}{2}\right) \overrightarrow{I_{\mathrm{DStatcom}}} \tag{2.16}$$

$$Q_\mathrm{D} = Im\left(V_{n+1} \angle \theta_{n+1} \left(\mathrm{I}_D \angle \left(\theta_{n+1} + \frac{\pi}{2}\right)\right)^*\right) \tag{2.17}$$

The cost of D-STATCOM is calculated as follows:

$$C_{\mathrm{DST}} = K_{\mathrm{DS}} \times Q_{\mathrm{DS}} \times \frac{(1+B)^{\mathrm{ND}} \times B}{(1+B)^{\mathrm{ND}} - 1} \tag{2.18}$$

where $Q_{\mathrm{DS}}$ is the rated of the D-STATCOM in kVar, $K_{\mathrm{DS}}$ denotes the capital cost of D-STATCOM in $/kVAR, ND is the lifetime of the D-STATCOM in years, and $B$ is the asset rate of return of the D-STATCOM.

## 2.8 Photovoltaic Energy Systems

A stand-alone inverter is used in off-grid applications with battery storage. With backup diesel generators such as (PV–diesel hybrid power systems), the inverters may have additional control functions such as operating in parallel with diesel generators and bidirectional operation (battery charging and inverting). Grid-interactive inverters must follow the voltage and frequency characteristics of the utility-generated power presented on the distribution line. For both types of inverters, conversion efficiency is a very important consideration. Photovoltaic power systems can be classified as follows [77]:

- Stand-alone
- Hybrid
- Grid connected

### 2.8.1 Solar Photovoltaic

Solar PV technologies use some of the properties of semiconductors to directly convert sunlight into electricity. The advantages are that these technologies are characterized by zero emissions, silent operation, and a long-life service [78, 79]. They also require low maintenance and no fuel costs. In addition, solar energy is redundant and inexhaustible. However, it is weather-dependent, intermittent, and unavailable during the night. Given a high PV penetration level together with demand variations, power distribution systems would experience power fluctuations along with unexpected voltage rise, high losses, and low-voltage stability. Another drawback is that PV technologies require a high-capital investment cost [79].

The basic element of the PV conversion system is the PV cell, which is a just simple P-N junction. The equivalent circuit of the PV cell based on the well-known single-diode model is shown in Fig. 2.5. It includes the current source (photocurrent), a diode ($D$), series resistance ($R_s$) that describes the internal resistance to flow of current, and parallel resistance ($R_{sh}$) that represents the leakage current. The current-voltage (I-V) characteristics of the PV cell can be expressed as follows [45, 80, 81] (Fig. 2.6):

$$I = I_{ph} - I_s \left\{ \exp\left[\frac{q(V + IR_s)}{\text{AKT}}\right] - 1 \right\} - \left(\frac{V + IR_s}{R_{sh}}\right) \tag{2.19}$$

The light generated current ($I_{ph}$) mainly depends on the solar irradiance and working temperature of PV cell, which is expressed as follows [82, 83]:

$$I_{ph} = [I_{sc} + K_i(T - T_{ref})].\left(\frac{G}{1000}\right) \tag{2.20}$$

The PV saturation current ($I_s$) varies as a cubic function of the PV cell temperature ($T$), and it can be described as follows:

$$I_s = I_{rs}\left(\frac{T}{T_{ref}}\right)^3 \exp\left[\frac{qE_g}{KA}.\left(\frac{1}{T_{ref}} - \frac{1}{T}\right)\right] \tag{2.21}$$

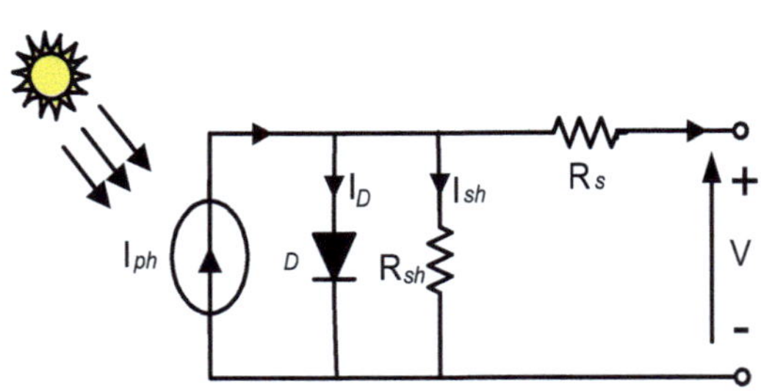


**Fig. 2.5** Equivalent circuit of the PV cell

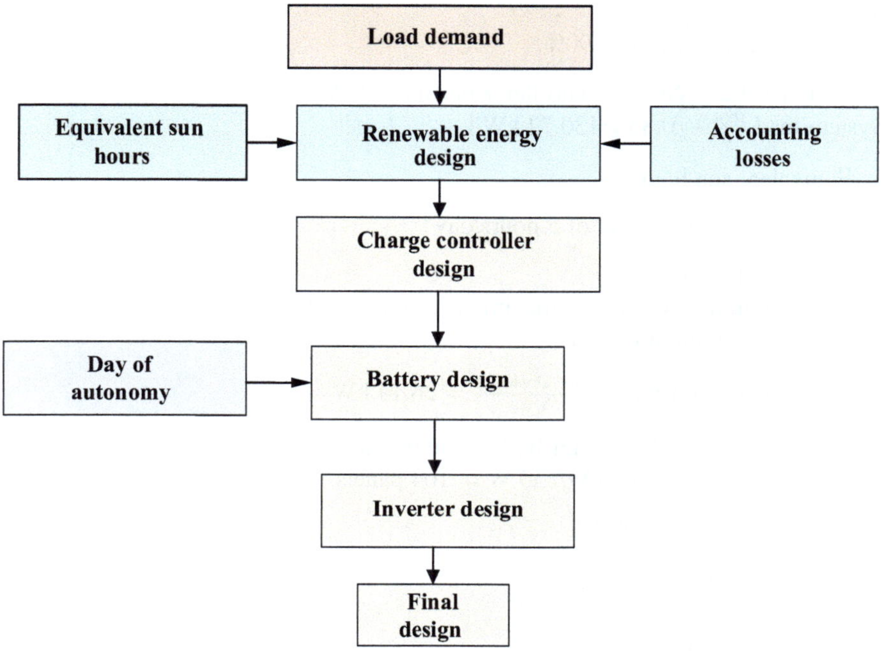


**Fig. 2.6** A flowchart to design a stand-alone system

The reverse saturation current ($I_{rs}$) can be approximately obtained as follows:

$$I_{rs} = \frac{I_{sc}}{\left[\exp\left(\frac{qV_{oc}}{N_{ser}.K.A.T}\right) - 1\right]} \tag{2.22}$$

where

$I$ and V are output current and voltage of PV cell respectively,
$I_s$ is the reverse saturation current of diode, and
$I_{ph}$ is the current generated by the incidence of light.
$R_s$ and $R_{sh}$ are series and parallel resistances of PV cell, respectively,
The photocurrent ,$I_{ph}$, is a function of temperature and solar insolation.

**Example 2.1**
A stand-alone system using PV array, 250 Watt and parameters are

1. Load demand
   Load demand = 100 kW
2. Day of autonomy
   Day of autonomy = 2 days of battery

3. Losses
   Inverter efficiency = 90%

Efficiency cabal + charge controller + battery = 85%
System load $\frac{100\text{kW}}{0.9}$ /0.85 =130.72 kWh

Equivalent sun hours

Rate of Sohag City the sun of 5 hours/day

PV array design

Find the number of panels, maximum current parallel, charge controller, battery capacity, and minimum Cbatt?

***Solution*** Minimum $Wp = \frac{130.72\text{kwh/day}}{5\text{h/day}} = 26.14$ kW

Number of panels = Minimum *Wp*/capability one
Number of panels = 26.4 kW/250 W = 104 panels

Maximum current parallel

*I*max = 8.52 × 104 = 886.08 A
*V*max = 37.22 × 104 = 3870.88 V

Charge controller

Operation voltage = 12 or 24 V
Depth of charge = 60%
Battery voltage = 12 or 24 or 36 ....
Battery capacity = 450 Ah

Battery design

Minimum Cbatt = total energy demand/depth of charge × operational voltage of the system × 2 days
Minimum Cbatt = $\frac{130.72\text{kwh}}{0.6\times 24} \times 2 = 18155.56$ Ah
Number of batteries in series = $\frac{24v}{12v} = 2$ batteries
Number of battery in parallel = 18155.6 Ah/450 Ah$\frac{18155.6\text{Ah}}{200\text{Ah}} = 40$ batteries
Number of batteries = 2 × 40 = 80 batteries

Inverter design

Minimum nominal power retting
100 kW/0.9 = 111 kW

## 2.8.2 Grid-Connected PV System

In a PV system, solar radiation is trapped and converted into electric power. PV cells are exposed to sunlight, the interior construction of PV cells can absorb photons, and then electrons are released. PV systems can be either on-grid or off-grid [84, 85]. The electric current by PV cells is always direct current (DC) in nature. It can be converted to alternative current (AC) by an electronic device inverter. There are four different topologies of PV inverter to be connected to the grid: (1) multi-string topology, (2) centralized topology, (3) string topology, and AC module topology [86]. It illustrated the PV system connection with the mesh power system. In this chapter, the PV-DG is modeled in a power system either PQ node or PV note. At the same time, it presents the modeling of the output power of the PV system based on solar irradiance [86] (Fig. 2.7).

**Example 2.2**
A grid-connected system using PV array 250 Watt and parameters are

Insolation = 1000 W/m$^2$
Efficiency PV = 20%
Efficiency inverter = 90%
Efficiency cable = 98%
Load demand = 100 kW

Equivalent sun hours

Rate of Sohag City the sun of 5 hours/day

Losses

System load $= \frac{100\ \text{kw}}{0.9}/0.98 = 113.37$ kWh

PV array design

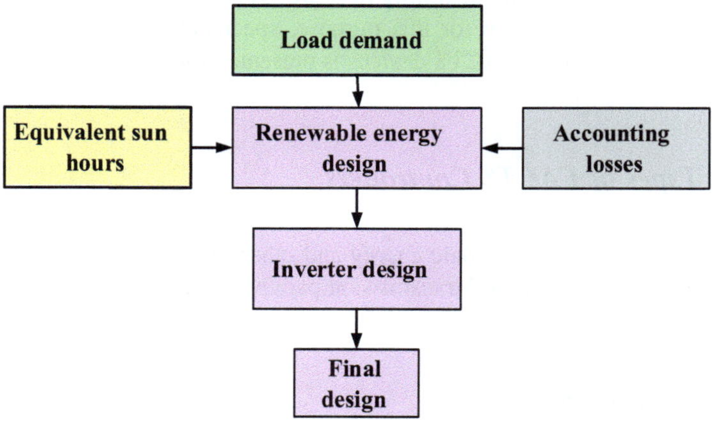


**Fig. 2.7** A flowchart to design a grid-connected system

Find the number of panels, maximum current parallel, charge controller, battery capacity, and minimum Cbatt?

***Solution*** Minimum $Wp = \frac{113.37\text{kwh/day}}{5\text{h/day}} = 22.67\ \text{kW}$

Number of panels = Minimum *Wp*/capability one
Number of panels $= \frac{22.67\text{kw}}{250\text{w}} = 100$ panels

Maximum current parallel

$I\text{max} = 8.52 \times 100 = 852\ \text{A}$
$V\text{max} = 37.22 \times 100 = 3722\ \text{V}$

Inverter design

Minimum nominal power retting
100 kW/0.9 = 111 kW

## 2.9 FACTS Devices

A flexible AC transmission system refers to a system consisting of power electronic devices along with power system devices to enhance the control and stability of the transmission system and increase power transfer capabilities. With the invention of the thyristor switch, the door was opened to develop of power electronics devices known as (FACTS) controllers. Basically, the FACTS system is used to provide the ability to control the high-voltage side of the network by integrating power electronic devices to provide inductive or capacitive power in the network [89].

The development of FACTS in the power transmission system has led to many applications of these controllers not only to improve various stability problems but also to provide the operating flexibility of power systems. There are different types of FACTS devices available for this purpose, specifically in the following subsections; concept of these FACTS devices is presented [90].

### *2.9.1 Types of FACTS Controllers*

The system is based on electronic energy and other fixed equipment that provides control one or more of the AC transmission parameters. FACTS controllers can be categorized as:

- Shunt controller connected
- Series connected controller

- Combined series of controllers
- Combined shunt-series controllers

### 2.9.2 Shunt Compensation

The reactive current is injected into the line to maintain the voltage magnitude. The active power of the transmitter is increased, but you must provide more reactive power as described in Fig. 2.8.

### 2.9.3 Series Compensation

FACTS to series compensation modify line impedance: X is reduced to increase the active power that can be transferred. However, a more reactive power force should be provided in Fig. 2.9.

### 2.9.4 PV-DG Operation Modes

Depending on the type, electric machine of DG, and utility interface (direct or indirect) used for a connection to the power system, all DGs have to be models either as PQ (power-controlled) node or PV (voltage-controlled) node power flow

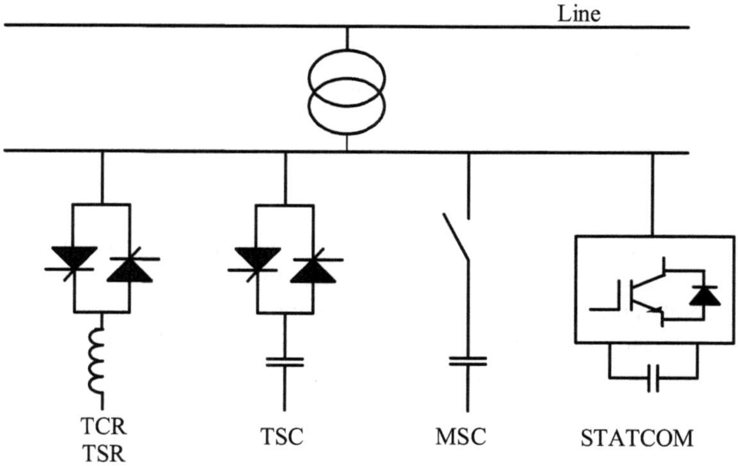


**Fig. 2.8** Examples of FACTS to shunt switcher

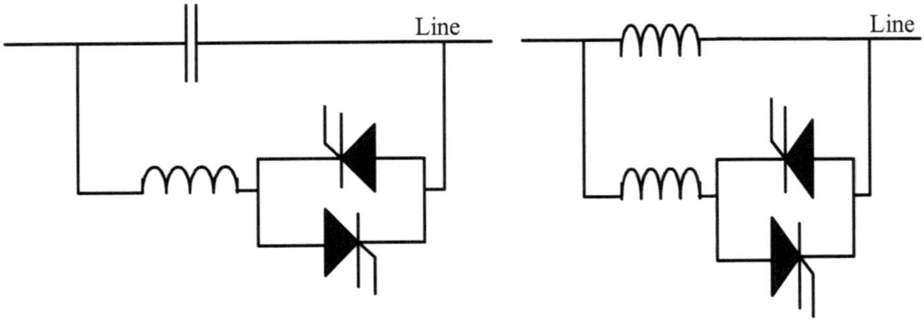


**Fig. 2.9** Examples of FACTS to series switcher

studies [91]. PV-DG is installed to the power system via a power electronic interface. Based on the control mode of the electronic device, the PV-DG model is determined.

In the case of controlling P and V independently, the PV DG model is set to PV node. On other hand, the PV-DG model is a PQ node when P and Q are controlled independently [92]. In the case of the PQ node, the PV-DG behaves as a negative load which does not make any obstacle to power flow. Meanwhile, PV-DG can operate as the PV node which requires an iterative procedure to keep specified voltage magnitude and monitor DG reactive power limits. Therefore, some means should be considered [93–95].

## 2.10 D-FACTS Devices

Depending on the type, electric machine of DG, and utility interface (direct or indirect) used for a connection to the power system, all distributed flexible AC transmission system (D-FACTS) devices or "smart wires" change the effective line impedance of the transmission line on which they are installed. The simulator supports the response of D-FACTS devices responds based on online current. This functionality is described in [96]. The ability to effectively control power flow in a network can allow better utilization of the existing network by routing power flow away from overloaded lines. The problem of power flow control is even more compelling when considering the continually changing network topology and the need to remain secure under contingency conditions. Also, studies indicate that transient stability and dynamic stability can be improved when reactive compensation is available and can be varied rapidly by electronic control [97].

System benefits resulting from power flow control have been the motivating factors for the use of flexible AC transmission system (FACTS) devices since the time of their development. More recently, distributed flexible AC transmission system (D-FACTS) devices [96] such as the distributed static series compensator (DSSC) have been designed to address power control types of problems. D-FACTS devices attach directly to transmission lines and can be used to dynamically control effective line impedance. Also, D-FACTS devices are smaller and less expensive than traditional FACTS devices which may make them better candidates for wide-scale deployment [98].

The FACTS devices are used for maintaining, either supporting or preventing, raising the voltage by supplying or absorbing the reactive power, but not exclusively for the improvement of the quality of the supply. The D-FACTS devices are exclusively for the improvement of the quality of the supply. Widespread use of conventional FACTS controllers has not extensively occurred due in part to size, expense, and installation effort. The use of D-FACTS devices may facilitate the realization of a comprehensively controllable power system. Large-scale power flow control may finally be achievable [98].

A D-FACTS device changes the effective line impedance actively by producing a voltage drop across the line which is in quadrature with the line current. Thus, a D-FACTS device provides either purely reactive or purely capacitive compensation. D-FACTS devices do not change the line's resistance at all since doing so would imply the ability of the device to create real power. The impact on the system caused by D-FACTS devices on different lines working together can be coordinated to achieve some desired control objective. D-FACTS devices may be configured to operate autonomously in certain situations such as during transients, faults, etc. [98–101].

Distributed FACTS devices:

- Capacitive or inductive
- Distributed static series compensator (DSSC)
- Distributed series reactor (DSR)

Synchronous voltage source improved operation of distribution networks with the use of D-FACTS device.

## 2.11 Integration of DGs with Distribution Networks

Green energy sources, such as hydroelectric power, wind turbine and solar energy do not emit smoke or create pollution when they are used [3].

With the increasing efforts toward electricity generation from green energy sources, there is a need to integrate the distributed energy resources with the power system, which may be integrated with the placement of DGs and D-FACTS devices at the suitable locations [100–104].

## 2.12 Power Quality

The design of electrical power supply systems is a compromise between the interests of consumers – reliability and quality of supply – and those of the supply industry – realistic investment levels and operating costs. The flexibility allowed to deviate from "perfect" power quality should be used to allow cheaper and simpler supply systems; it should not be wasted by permitting poor maintenance and operating procedures to compromise reliability [105].

Consider electric energy as a product: it is generated, transmitted, distributed, and sold to customers. The end-user converts the electric energy into other forms such as mechanical, thermal, and light energy. The users of electric energy expect a reasonable degree of reliability and quality of service. In technical terms, the following conditions are required to ensure customer satisfaction [106].

The electric energy must be continuously available (reliable supply).

The voltage supply must alternate at a constant frequency with a sinusoidal waveform and a constant magnitude.

The voltage magnitude must be within the range recommended by the equipment manufacturer.

In three-phase systems, there must be perfect symmetry: The three voltages must be identical sinusoids shifted 1,200 with respect to each other.

## 2.13 Power Quality Disturbances Classification

The effects of power disturbances vary from one piece of equipment to another and with the age of the equipment. Equipment that is old and has been subjected to harmful disturbances over a prolonged period is more susceptible to failure than new equipment. To classify different types of power quality disturbances, the characteristics of each type must be known. In general, power quality disturbances are classified into two types: steady-state and non-steady-state. This classification is done in terms of the frequency components which appear in the voltage signals during the disturbance, the duration of the disturbance, and the typical voltage magnitude. These disturbances are mainly caused by [107]:

- External factors to the power system: for example, lightning strikes cause impulsive transients of large magnitude.
- Switching actions in the system: a typical example is capacitor switching, which causes oscillatory transients.
- Faults which can be caused, for example, by lightning (on overhead lines) or insulation failure (in cables).
- Voltage dips and interruptions are disturbances related to faults.
- Loads which use power electronics and introduce harmonics to the network.

## 2.14 Power Quality Issues

A recent study of power quality experts indicates that 50% of all power quality issues are related to grounding, earth bonds, neutral to ground voltages with earth loops, ground current, or other ground-associated issues. Equipment that is powered by electricity or related to it is affected by the power quality. Identification of the issues requires sophisticated electronic testing equipment. The following symptoms are indicators of power quality issues [108]:

- A piece of equipment is malfunctioning at the same time of day.
- Trip circuit breakers without being overloaded.
- Equipment failure during thunderstorms.
- Automated systems shall cease for no apparent reason [109].

In the power system, the power quality indicates a high degree of electrical service. The need for end-user equipment determines the desired power quality. In general, power quality refers to the maintenance of a sinusoidal waveform for voltage and frequency voltages, the concept of powering, and the grounding of sensitive equipment in an appropriate manner to operate the equipment. Some common power quality problems include voltage drops (Sag), voltage amplification, interruptions, transient or voltage flicker, voltage unbalancing, harmonics, DC offset, notches, and noise, etc. [110].

The problems of power quality have increased significantly according to power electronic-based devices in the electrical distribution system. D-STATCOM is a promising shunt-connected custom device designed to mitigate power quality problems. Investigation on the SRF-based D-STATCOM system is performed with the grid-connected PV system using a fuzzy logic controller via the PI controller for reactive power management. Distribution static synchronous compensator (D-STATCOM) is an inverter-based power-quality conditioner connected in shunt with an AC system. It is used to correct power factor, load balancing, voltage regulation, and harmonic filtering in distribution systems [111].

The distribution system is characterized by poor power quality due to the lack of sufficient reactive energy during the state of stability and dynamism. Because of the extensive use of automatic, this is addressed by modeling and analyzing dedicated power controllers and electronic power-based equipment designed to enhance the reliability and quality of power flow in low-voltage distribution networks using D-STATCOM. A new PWM-based control system that requires only voltage measurements is operation, and the proposed D-STATCOM control mode is displayed [112]. This proposes the power quality improvement of stand-alone hybrid power generation employing solar power and wind power using FACTS devices. The hybrid power generation system, which uses natural energy, will integrate the protection of the global environment into isolated and rural areas without any dependence on the commercial power system. This shows the use of the FACTS device toward improving the power quality by removing the harmonics in the voltage [113].

## 2.15 Classification of Power Quality Disturbances

Power quality phenomena classification and characterization are shown in the chart of Fig. 2.10.

### 2.15.1 Transient Power Quality Disturbance

Transient PQ disturbances are momentary transient disturbances are classified to.

Impulsive transient PQ disturbances (Fig. 2.11) which are sudden, non-power frequency change in the steady-state voltage or current waveforms with unidirectional polarities characterized by rising and decay times mainly caused by lightning.

Oscillatory transient PQ disturbances (Fig. 2.12) are sudden, non-power frequency change in the steady-state condition of voltage, current, or both, including positive and negative values, characterized by its spectral contents, duration, and

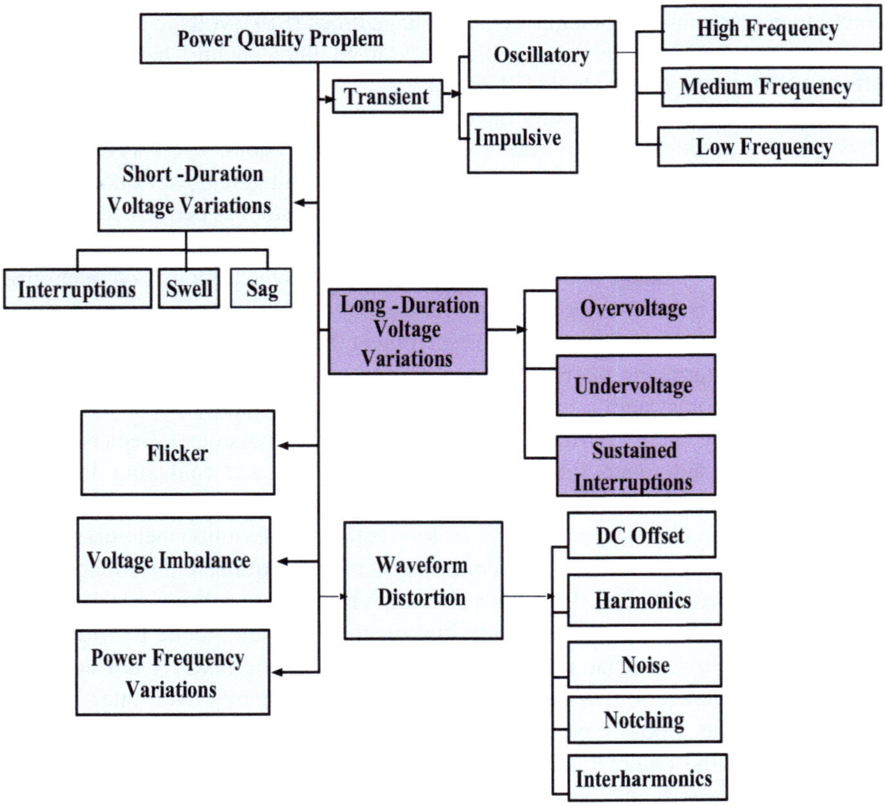


**Fig. 2.10** Power quality disturbances

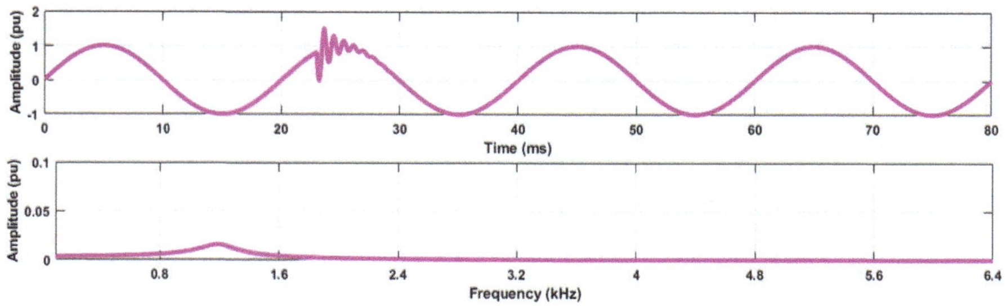


**Fig. 2.11** Impulsive transient PQ disturbance

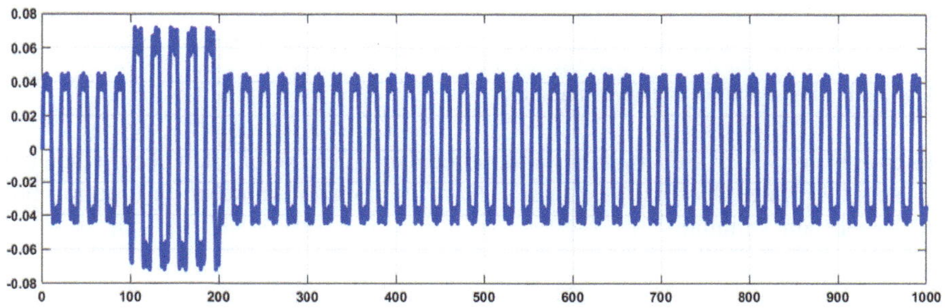


**Fig. 2.12** Oscillatory transient PQ disturbance

magnitude [113]. Based on spectral contents, oscillatory transients can be classified into high-, medium-, and low-frequency oscillatory transients. The oscillatory transient may be caused by power electronic commutation (several kHz), capacitor bank energization on distribution system (low frequency).

### 2.15.2 *Short-Duration Voltage Variations*

Depending on duration of voltage variations, short-duration voltage disturbances, Fig. 2.13, can be classified into three categories:

Voltage sag: this is a drop in voltage (between 10% and 90%) for a few cycles (1–30). This is caused by a fault in the power system in remote locations. Voltage sag can also be the result of a lightning strike that operates a circuit breaker or a reclosure, which recloses instantaneously (3–10 cycles). Also, motors – both constant-speed and ASD – may cause voltage sag (for about 65% voltages) [113].

Voltage swells: this is an increase in the voltage between 1.1 and 1.8 p.u. at the power frequency for a short period (1–30 cycles). The cause can be changed in system loading (a large load is dropped), switching on of a capacitor bank, etc. The lights blink as an indication of swells.

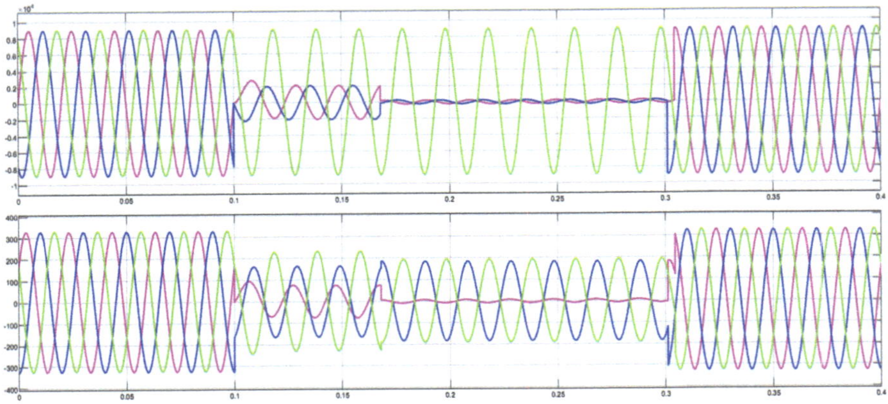

Fig. 2.13 Short-duration voltage disturbances

Interruptions: which are temporary and momentary interruptions that occur when the supply voltage decrease less than 10% of its nominal value up to a period not larger than 1 minute.

### 2.15.3 Long-Duration Voltage Variations

Duration of long-duration voltage disturbances is generally larger than 1 minute. These PQ disturbances can be classified to:

Undervoltage: is a drop in voltage below the standard voltage tolerance (±5%). Most equipment will operate in the range of ±10% for a short period without damage. Starting large motors may cause the voltage to drop close to 10% [108].

Overvoltage: is an increase in voltage above the standard voltage tolerance (±5%). Over voltages can be a result of erroneous system operation, line-to-ground faults, or switching of major system components.

Sustained interruptions: which are interruptions that occur when the supply voltage decreases less than 10% of its nominal value for a period larger than 1 minute [108].

Voltage imbalance: voltage imbalance is defined as the ratio of negative sequence voltage (and/or the zero-sequence voltage) to the positive sequence voltage. Voltage imbalance can result from unbalanced loads [113].

## 2.15.4 Waveform Distortion

The steady-state deviation of voltage and/or current waveforms from the pure sinusoidal power frequency waveforms is referred to as waveform distortion phenomena in Fig. 2.14. Waveform distortion PQ problem can be classified [114].

DC offset: is defined by the presence of a DC bias in the voltage and/or current waveforms. DC offset results from the operation of rectification equipment such as half-wave rectifiers. DC offset can cause saturation problems in transformers and machines [114].

Harmonics: are sinusoidal voltages and/or currents of frequencies that multiples of the fundamental system frequencies. The use of converters, nonlinear loads, etc. is the major source of harmonics [49].

Interharmonics: which are harmonics with frequencies that are not integer multiples of the fundamental frequency. Interharmonics can be caused by static frequency converters, cycle converters, and arcing devices.

Notching: is a periodic voltage disturbance caused by the normal operation of power electronic devices as the current is commutated from one phase to another Fig. 2.15 [49].

Electrical noise: is the distortion (not necessarily periodic) with broadband spectral contents lower than 200 kHz superimposed on the power system voltage and/or current waveforms. Switching radio transmitter and arcing industrial equipment can cause electrical noise [113].

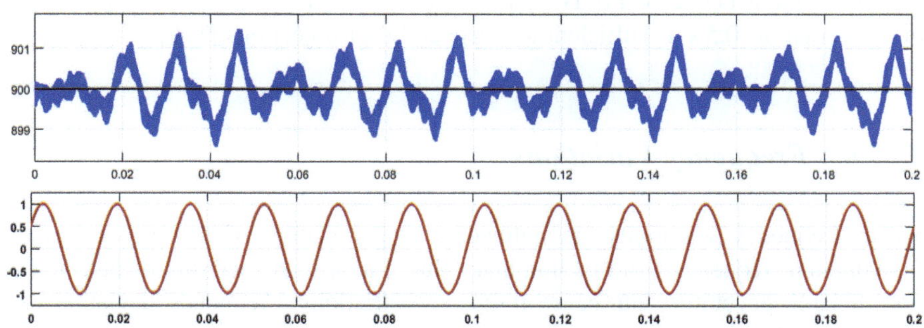


**Fig. 2.14** Waveform distortion

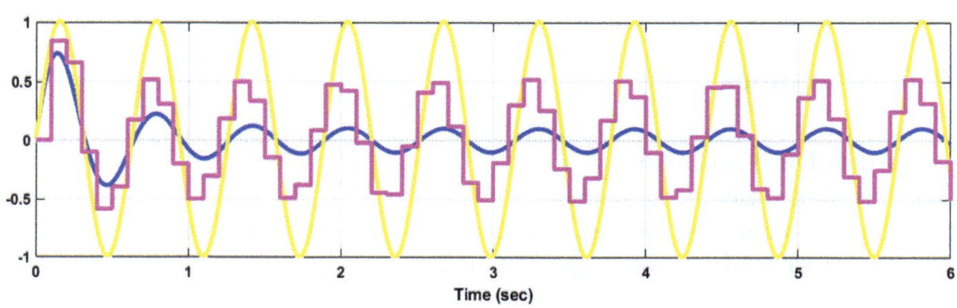


**Fig. 2.15** Commutation notches

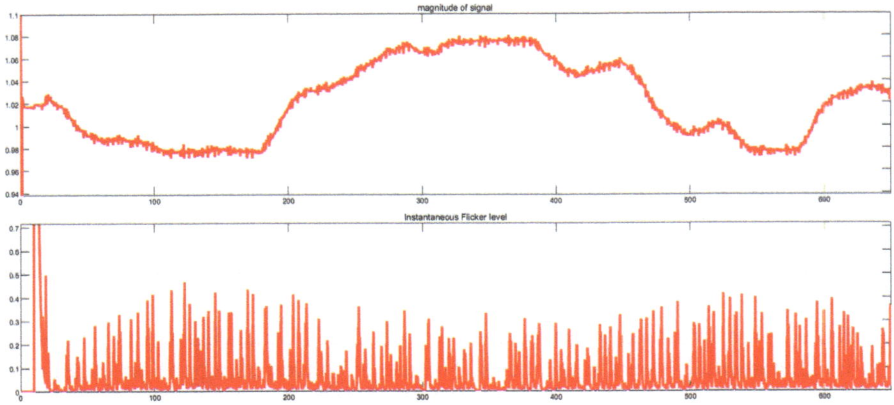


**Fig. 2.16** Voltage fluctuations

### 2.15.5 *Flicker*

Flicker is the term used to define the fluctuations of voltage, Fig. 2.16, caused by loads that vary in frequency. Flicker is the perceptible change in the output of a source of light when there is a sudden change in the supply voltage (sag or swell). The voltage fluctuations cause light pulsations. If such disturbances are recurrent and the incremental change in the luminous flux is large enough, the flicker becomes annoying [115]. The eye–brain perception system has maximum sensitivity at about nine fluctuations per second (9 Hz). At this critical frequency, a voltage dip or swell of V as small as 0.5% is sufficient to cause end-user objections [115].

### 2.15.6 *Frequency Variations*

Power frequency variations are defined as the deviation of the power system fundamental frequency from its nominal value. Large load shedding or disconnection of large generators can cause frequency variations. Frequency deviation can affect the performance of electronic timers and also can reduce measuring equipment accuracy [115].

## 2.16 Questions

Define distributed generation?
Draw distributed electricity systems?
Mention the commercial and industrial sectors, distributed generation?
What are distributed generation applications?

What are distributed generation technologies?
Define distributed static compensator (D-STATCOM)?
What is the power quality impact of PV-DG?
Draw the schematic diagram of the D-STATCOM device?
Classify photovoltaic power systems?
What are the facts devices?
Mention the types of FACTS controllers?
Draw the examples of facts to shunt switcher?
Mention the D-FACTS devices?
Why there are power quality issues?
What is power quality?
Classify power quality disturbances?

# Chapter 3
# Stochastic Optimal Planning of Distribution System Considering Integrated Photovoltaic-Based DG and D-STATCOM

## 3.1 Distributed Network

The electric distributed network mainly delivers the electric power from the high-voltage transmission system to the consumers. In electric distribution networks, the R/X ratio is significantly high compared to transmission systems; hence, power loss is high (about 10–13% of the generated power). Moreover, poor quality of power including the voltage profile and voltage stability issues may arise. Therefore, the power quality of the service is based on the continuity of power and maintaining the supply voltage within certain limits with a specified frequency. Radial distribution system (RDS) includes various loads such as commercial, industrial, residential, etc. [136].

The inclusion of shunt capacitors and distributed flexible AC transmission system (D-FACTS) devices can significantly enhance the performance of electric distribution networks by providing the required reactive power compensation. D-FACTS include different members such as distributed static compensator (D-STATCOM), photovoltaic-distributed generation (PV-DG). Optimal allocation of these controllers in the electric distribution networks is a strenuous task for researchers for power loss minimizing, voltage profile improvement, voltage stability enhancement, reducing the system overall costs, and maximizing the system load ability and reliability.

## 3.2 Backward/Forward Sweep (BFS) Algorithm

The BFS algorithm is one of the most common methods used for load flow distribution systems because it is simple, fast, and robust in convergence and has a low memory requirement for processing with high efficiencies and solution accuracy. The BFS algorithm involves mainly three basic steps based on Kirchhoff's current law (*KCL*) and Kirchhoff's voltage law (*KVL*). The three steps are named the

 
A. M. A. El-Hameid et al., *Enhancement of Grid-Connected Photovoltaic Systems Using Artificial Intelligence*, https://doi.org/10.1007/978-3-031-29692-5_3

nodal current calculation, the backward sweep, and the forward sweep, and they are repeated until the convergence is achieved. The BFS utilizes a simple and flexible radial distribution system numbering scheme to number each branch in the feeder, lateral, and sublateral [136]. The BFS algorithm can be applied to find the load flow results using the following steps:

**Step 1: Initialization which include the following:**

- The distribution system line and load data.
- The base power and base voltage.
- Calculate the base impedance.
- Calculate the per-unit values of line and load data.
- Take the voltage for all bus's float voltage (1 p.u.).
- Set convergence tolerance $\epsilon = 0.0001$ and $\Delta V_{\max} = 0$.

**Step 2: Radial Distribution System Numbering Scheme**

The numbering scheme aims to give a number to each section in the distribution system, where a section is a part of a feeder, lateral, or sublateral that connects two buses in the distribution system. The total number of sections $N_{\text{Sec}}^{\text{Total}}$ of a distribution system can be calculated as:

$$N_{\text{Sec}}^{\text{Total}} = N_{\text{bus}}^{\text{Total}} - 1 \tag{3.1}$$

where $N_{\text{bus}}^{\text{Total}}$ is the total number of buses. Each section will carry a number which is one less than its receiving end bus number; for example, the number of sections that connects the sending end $p$ and the receiving end $q$ in Fig. 3.1 can be calculated as:

$$N_{\text{Sec}/P-q} = N_{\text{bus}/q} - 1 \tag{3.2}$$

where $N_{\text{Sec}/P-q}$ is the section number between buses $p$ and $q$ and $N_{\text{bus}/q}$ is the number of bus $q$. Now, the radial distribution system numbering scheme should be applied on the distribution system to give a number to each section in the system.

**Step 3: Nodal Current Calculation**

At iteration $k$, the nodal current injection at node $i$ due to loads and any other shunt elements can be calculated as:

$$I_i^k = \left(\frac{S_I}{V_i^{(k-1)}}\right) - (Y_i)\left(V_i^{(k-1)}\right) \tag{3.3}$$

where $I_i^k$ is the current injection at node $I$, $S_I$ is the specified power injection at node $I$, $V_i^{(k-1)}$ is the voltage at node $I$ at iteration $k-1$, and $Y_i$ is the sum of all shunt elements at node $i$.

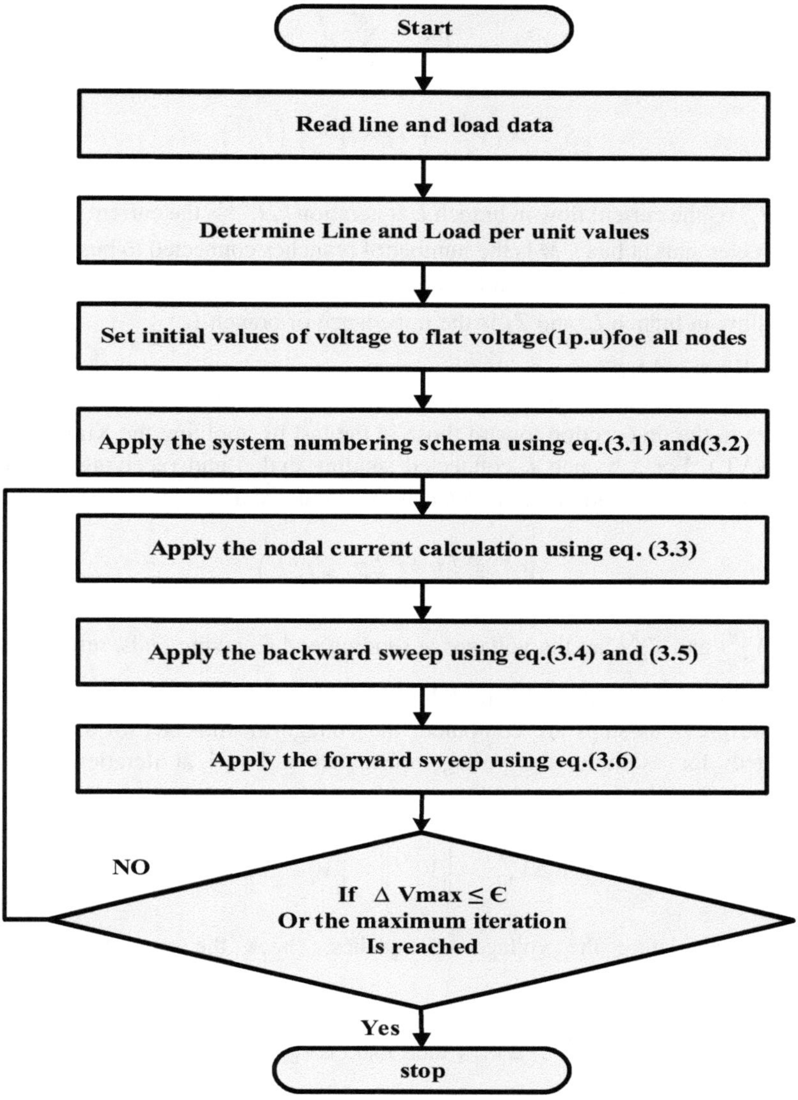


**Fig. 3.1** Flow chart of BFS load flow

### Step 4: Backward Sweep

At iteration *k*, start from the branches at the end nodes and move toward the branches connected to the substation. Hence, all branch currents can be calculated by applying the Kirchhoff Current Law (*KCL* (and then the powers through these branches can be determined as:

$$I_L^{(K)} = -I_j^{(K)} - \sum_{m=1}^{M} \left( \frac{S_m}{V_j^{(K)}} \right) \tag{3.4}$$

$$S_L^{(K)} = \left( V_j^{(K)} + Z_L \times I_L^{(K)} \right) \left( I_L^{(K)} \right) \tag{3.5}$$

where $I_L^{(K)}$ is the current flow in branch $L$ at iteration $k$, $I_j^{(K)}$ is the current injected due to shunt elements at bus $j$, $M$ is the number of branches connected to bus $j$, Sm is the complex power at the sending end of branch m, $V_j^{(K)}$is the voltage at bus $j$, $S_L^{(K)}$is the power flow in branch $L$, and $Z_L$ is the impedance of branch $L$.

**Step 5: Forward Sweep**
At iteration $k$, the nodal voltages are updated in a forward sweep starting from the branches in the first section toward those in the last by applying the Kirchhoff Volt Law) KVL). For a branch $L$ connected sending end $p$ and receiving end $q$, the voltage at receiving end at iteration $k$ can be calculated as:

$$V_q^{(K)} = \left( V_p^{(K)} - Z_L * I_L^{(K)} \right) \tag{3.6}$$

where $V_p^{(K)}$ and $V_q^{(K)}$ are the voltages at sending and receiving ends, respectively.

**Step 6: Check the Voltage Mismatches**
After the previous steps are computed, the voltage mismatches for all nodes are calculated; for example, the voltage mismatch at bus i at iteration k can be calculated as:

$$\Delta V_i^{(K)} = \left\| V_i^{(K)} \right\| - \left\| V_i^{(K-1)} \right\| \tag{3.7}$$

After calculating the voltage mismatches, check the convergence of the voltage as:

$$\Delta V_i^{(K)} > \Delta V_{\max} \text{ then make} \Delta V_{\max} = \Delta V_i^{(K)} \tag{3.8}$$

If $\Delta V_{\max} \geq \in$, go to step 8, otherwise increment the iteration number, and go to step 3.

**Step 7: Check Stopping Criterion**
The program will be terminated when the maximum iteration is reached or the convergence from the voltage mismatches is verified.

## 3.3 Forward/Backward Power Flow

The main feature of the radial distribution network is that it has a high R/X ratio. Thus, the conventional power flow method cannot solve the power flow for these networks because these methods suffer from low convergence characteristics. As a result, the forward/backward (FB) load flow method is more applicable than the Newton–Raphael method for solving power flow in distribution grids. The power of forward and backward depends upon the power of Kirchhoff's current and voltage laws. Fig. 3.2 depicts a simple radial network. By applying Kirchhoff's law to the depicted Fig. 3.2, the following equations can be obtained, which describe the forward–backward method.

$$\bar{I}_L = \left( \frac{\bar{S}_{L,n}}{\bar{V}_n} \right)^* \tag{3.9}$$

where $\overline{S_{L,n}}$, $\bar{I}_L$, and $\bar{V}_n$ denote the apparent power current, and voltage, *t* bus n, respectively. The backward step of the FB method depends upon calculating the branch current as follows:

$$\bar{I}_n = \bar{I}_k + \sum_{p \in M} \bar{I}_p \tag{3.10}$$

The forward step is based on calculating and updating the buses voltage as follows:

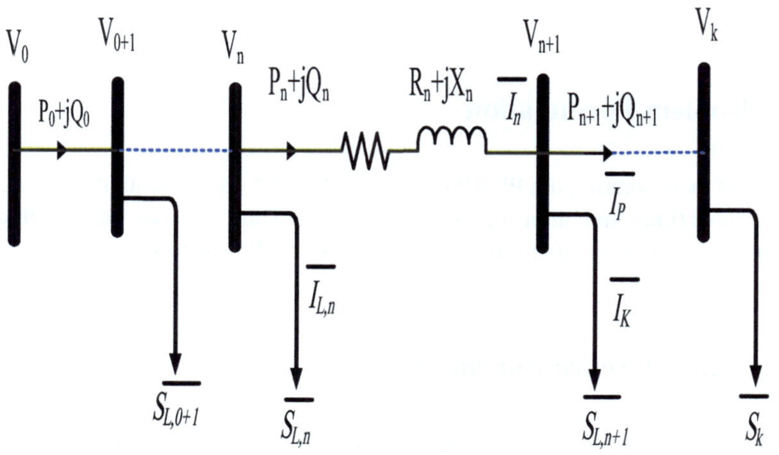


**Fig. 3.2** A radial distribution network

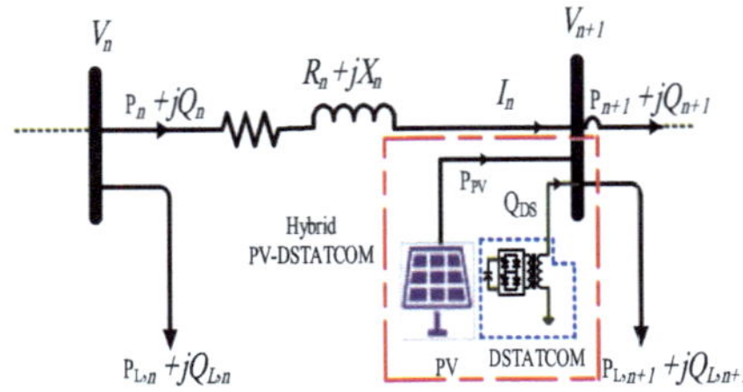


**Fig. 3.3** The system with the PV and D-STATCOM

$$\overline{V_{n+1}} = \overline{V_n} - (R_n + jX_n)\overline{I_n} \tag{3.11}$$

In the case of the inclusion of the hybrid PV-DG and D-STATCOM as depicted in Fig. 3.3, the power flow is expressed as follows:

$$P_{n+1} = P_n - P_{L,n+1} - R_{n,n+1}\left(\frac{P_n^2 + jQ_n^2}{|V_n|^2}\right) + P_{\text{PV}} \tag{3.12}$$

$$Q_{n+1} = Q_n - Q_{L,n+1} - X_{n,n+1}\left(\frac{P_n^2 + jQ_n^2}{|V_n|^2}\right) + Q_{\text{DS}} \tag{3.13}$$

where $X$ and $R$ denote the reactance and resistance of the transmission line, respectively. $Q_{\text{Ds}}$ is the reactive power injected by D-STATCOM, while $P_{\text{PV}}$ is the active power inserted by PV unit. The active power loss of line $n$, $n + 1$ is calculated as follows:

$$P_{\text{Loss}} = R_{n,n+1}\left(\frac{P_n^2 + jQ_n^2}{|V_n|^2}\right) \tag{3.14}$$

## 3.4 Problem Formulation

The aim of incorporating the PV-DG and D-STATCOM is reducing the annual cost, voltage deviations and improve system stability. Thus, the considered objective function is a multi-objective function which described as follows:

### *3.4.1 The Objective Functions*

This book presents a single objective function and multi-objective function that can be discussed as follows:

#### 3.4.1.1 Single Objective Function

The studied single objective function is the total active power loss of the system and can be formulated as follows:

$$F_{\text{obj}} = \sum\nolimits_{l=1}^{nl} (P_{\text{loss}}(l)) \tag{3.15}$$

where *nl* is the total number of branches in RDS. The active power losses of branch *n* and *n* + 1 is given as follows:

$$P_{\text{loss}} = R_{n,n+1} \left( \frac{P_n^2 + jQ_n^2}{|V_n|^2} \right) \tag{3.16}$$

The total power loss equals to summation of power losses in all branches as follows:

$$P_{\text{Total\_loss}} = \sum_{i=1}^{\text{NT}} P_{\text{loss},i} \tag{3.17}$$

where NT is number of branches. The total active and reactive power losses of a radial distribution system can be written as:

$$\text{Total } P_{\text{Loss}} = \sum_{q=2}^{N_b} \sum_{k=1}^{N_b-1} \left( \frac{P_{\text{eff}/q}^2 + Q_{\text{eff}/q}^2}{V_q^2} \right) * R_k \tag{3.18}$$

$$\text{Total } Q_{\text{Loss}} = \sum_{q=2}^{N_b} \sum_{k=1}^{N_b-1} \left( \frac{P_{\text{eff}/q}^2 + Q_{\text{eff}/q}^2}{V_q^2} \right) * X_k \tag{3.19}$$

where $P_{\text{eff}/q}^2$ and $Q_{\text{eff}/q}^2$ are the total effective active and reactive power loads beyond the node *q*, respectively. *Nb* is the total number of system buses, while *Nb* − 1 refers to the total number of system branches.

##### 3.4.1.1.1 Voltage Stability Index

In this book, voltage stability index (VSI) is used to indicate the security level of distribution system. VSI determines the sensitivity of each bus system to voltage collapse by Eq. (3.20) [137]. If the bus has high value of VSI, the bus is more stable and the possibilities of voltage collapse at this bus is weak. The optimal allocation of DERs and capacitors in distribution network increase the value of VSI for each bus in RDS. Summation of voltage stability index is the total value of VSI for all buses in RDS [138].

$$\mathrm{VSI}_{(n+1)} = |V_n|^4 - 4(P_{mn+1}X_n - Q_{n+1}R_n)^2 - 4(P_{n+1}X_n + Q_{n+1}R_n)|V_n|^2 \quad (3.20)$$

$$\mathrm{VSI}_{(n+1)} = \left(\frac{I}{\mathrm{SI}n}\right), n = 2, 3 \ldots\ldots n \quad (3.21)$$

where $\mathrm{VSI}_{(n+1)}$ represents the VSI for bus ($r$) and $V_n$ represents the voltage at bus ($s$) while $X_n$ and $R_n$ represent the reactance and the resistance between bus ($s$) and bus ($r$). But $P_{mn+1}$ and $Q_{n+1}$ represent the active and reactive power injection from bus ($r$) to RDS. The maximum value of SIn gives minimum value of objective function (VSI). So, minimum values of objective function indicate improvement of voltage stability index. This can be accomplished by maximizing the total voltage stability index deviations according to:

$$\mathrm{TVSI} = 91.25 \times \sum_{i=1}^{\mathrm{Ns}} \sum_{h=1}^{24} \sum\nolimits_{n=1}^{\mathrm{NB}} \mathrm{VSI}_n \quad (3.22)$$

where

$$\mathrm{VSI}_n = |V_n|^4 - 4(P_{mn+1}X_n - Q_{n+1}R_n)^2 - 4(P_{n+1}X_n + Q_{n+1}R_n)|V_n|^2 \quad (3.23)$$

where VSI is the voltage stability index of the $n$th bus.

#### 3.4.1.1.2 Voltage Deviation

Bus system voltage is an indication to power equality and security indices of the power system. Therefore, any change in bus voltage affects the performance of power system that is calculated by the voltage deviation (VD). The voltage deviation of the system is defined as:

$$\mathrm{VD} = \sum\nolimits_{h=1}^{n} |V_n - 1| \quad (3.24)$$

where $n$ is number of system nodes and $V_n = 1$ p. u

$$\mathrm{TVD} = 91.25 \times \sum_{i=1}^{Ns} \sum_{h=1}^{24} \sum\nolimits_{n=1}^{NB} |V_n - 1| \quad (3.25)$$

where TVD denotes the total voltage deviations; $N_{\mathrm{B}}$ denotes the number of buses in the grid.

#### 3.4.1.1.3 The Total Annual Cost

In this book, the considered objective function is minimizing the total annual cost considering the seasonal and hourly variations of the load and the solar irradiance are changed. The year includes four seasons (winter, spring, summer, and fall). The load demand and irradiance are changed seasonally and hourly; this can represent by daily carve for each season (24 × Ns) where *Ns* equals to 4 which denotes to number of seasons per year and it is repeated 91.25 times to represent 1 year ($91.25 \times 24 \times Ns = 8760\ h$). The objective function includes the annual energy loss cost ($C_{\text{Loss}}$), the cost of the purchasing electric power from grid ($C_{\text{Grid}}$), PV units cost ($C_{\text{PV}}$), and D-STATCOM cost ($C_{\text{DST}}$), and it can be represented as follows:

$$C_{\text{PV\&DST}} = \min\ (C_{\text{Loss}} + C_{\text{Grid}} + C_{\text{PV}} + C_{\text{DST}}) \tag{3.26}$$

It should be pointed out here that the load demand is varied daily and seasonally per year. Thus, the cost of energy losses can be expressed as follows:

$$C_{\text{Loss}} = K_{\text{loss}} \times 91.25 \times \sum_{i=1}^{Ns} \sum_{h=1}^{24} \sum_{j=1}^{\text{NT}} P_{\text{Loss}(i,h,j)} \tag{3.27}$$

where $K_{\text{loss}}$ represents the cost of energy loss. *N*s is the number of seasons per year, which is 4. NT is the number of the branches of the grid. 91.25 denotes the number of days per season. The second term of (3.26), which represents the purchased energy from the grid, can be described as follows:

$$C_{\text{Grid}} = K_{\text{Grid}} \times 91.25 \times \sum_{i=1}^{Ns} \sum_{h=1}^{24} P_{\text{Grid}(i,h)} \tag{3.28}$$

where $K_{\text{Grid}}$ is the cost of the purchasing power in *\$/kW* $P_{\text{Grid}}$ is the captured power from the grid. The cost of the PV unit ($C_{\text{PV}}$) is divided into two costs including the fixed cost ($C_{\text{PV}}^{\text{Fixed}}$) and the variable cost ($C_{\text{PV}}^{\text{Variable}}$). These costs can be calculated as follows:

$$C_{\text{PV}}^{\text{Fixed}} = \text{CRF} \times C_{\text{PV}} \times P_{\text{pr}} \tag{3.29}$$

where $C_{\text{PV}}$ is the purchased cost of the PV unit in \$/kW; $P_{\text{pr}}$ is the rated power of the PV unit. CRF denotes the capital recovery factor, which can be founded as follows:

$$\text{CRF} = \frac{\alpha \times (1+\alpha)^{\text{NP}}}{(1+\alpha)^{S} - 1} \tag{3.30}$$

where $\alpha$ denotes the rate of interest on capital investing of the connected PV. NP is the lifetime of the PV-DG in years. The variable cost that related to the PV is associated with output kWh as follows:

$$C_{\mathrm{PV}}^{\mathrm{Variable}} = C_{\mathrm{O\&M}} \times \sum_{i=1}^{Ns} \sum_{h=1}^{24} P_{\mathrm{PV}(i,h)} \tag{3.31}$$

$C_{\mathrm{O\ \&\ M}}$ is the operation and maintenance costs which is associated with the output kWh of the PV unit $P_{\mathrm{PV}}$ represents the PV output power, which can be given as follows:

$$P_{\mathrm{PV}} = \begin{cases} P_{\mathrm{sr}}\left(\dfrac{{g_{\mathrm{s}}}^2}{G_{\mathrm{STD}} \times X_{\mathrm{c}}}\right) & \text{for } 0 < g_{\mathrm{s}} \leq X_{\mathrm{c}} \\ P_{\mathrm{sr}}\left(\dfrac{g_{\mathrm{s}}}{G_{\mathrm{STD}}}\right) & \text{for } X_{\mathrm{c}} \leq g_{\mathrm{s}} \leq G_{\mathrm{STD}} \\ P_{\mathrm{sr}} & g_{\mathrm{STD}} \leq g_{\mathrm{s}} \end{cases} \tag{3.32}$$

where $g_{\mathrm{s}}$ denotes the solar irradiance in W/ m$^2$. $G_{\mathrm{std}}$ denotes the solar irradiance in a familiar natural environment of about 1000 W/m$^2$. $X_{\mathrm{c}}$ is a certain irradiance point. The cost of D-STATCOM is calculated as follows:

$$C_{\mathrm{DST}} = K_{\mathrm{DS}} \times Q_{\mathrm{DS}} \times \frac{(1+B)^{\mathrm{ND}} \times B}{(1+B)^{\mathrm{ND}} - 1} \tag{3.33}$$

where $Q_{\mathrm{DS}}$ is the rated of the D-STATCOM in kVar, $K_{\mathrm{DS}}$ denotes the capital cost of D-STATCOM in \$/KVar, ND is the lifetime of the D-STATCOM in years, and $B$ is of the D-STATCOM.

#### 3.4.1.2 Two-Objective Function

This function can be formulated as follows:

$$F_1 = \frac{P_{\mathrm{T,loss}}}{(P_{\mathrm{T,loss}})_{\mathrm{base}}} \tag{3.34}$$

$$F_2 = \frac{\mathrm{VD}}{(\mathrm{VD})_{\mathrm{base}}} \tag{3.35}$$

$$F_3 = \frac{1}{\sum_{i=1}^{nb} |\mathrm{VSI}(i)|_{\mathrm{base}}} \tag{3.36}$$

The generalized objective function can be formulated as follows:

$$F = H_1F_1 + H_2F_2 + H_3F_3 \tag{3.37}$$

where $H_1$, $H_2$, and $H_3$ are weighting factors. Value of any weighting factor is selected based on the relative important on the related objective function with others objective functions. The sum of the absolute values of the weighting factors in Eq. (3.38) assigned to all impacts should add up to one as follows:

$$|H_1| + |H_2| + |H_3| = 1 \tag{3.38}$$

#### 3.4.1.2.1 Multi-Objective Function

This function can be formulated as follows:

$$F_1 = \frac{P_{\mathrm{T,loss}}}{(P_{\mathrm{T,loss}})_{\mathrm{base}}} \tag{3.39}$$

$$F_2 = \frac{\mathrm{VD}}{(\mathrm{VD})_{\mathrm{base}}} \tag{3.40}$$

$$F_3 = \frac{1}{\sum_{i=1}^{nb} |\mathrm{VSI}(i)|_{\mathrm{base}}} \tag{3.41}$$

$$F_4 = \frac{C_{\mathrm{PV\&DST}}}{\mathrm{Cost}_{\mathrm{base}}} \tag{3.42}$$

The generalized objective function can be formulated as follows:

$$F = H_1F_1 + H_2F_2 + H_3F_3 + H_4F_4 \tag{3.43}$$

where $H_1$, $H_2$, and $H_3$ are weighting factors. Value of any weighting factor is selected based on the relative important on the related objective function with others objective functions. The sum of the absolute values of the weighting factors in Eq. (3.44) assigned to all impacts should add up to one as follows:

$$|H_1| + |H_2| + |H_3| + |H_4| = 1 \tag{3.44}$$

#### 3.4.1.2.2 The Total Annual Cost

$$\min\ F(X,U) = \min(\text{Cost}_{\text{loss}} + \text{Cost}_{\text{Grid}} + \text{Cost}_{\text{PV}} + \text{Cost}_{\text{DS}}) \tag{3.45}$$

where

$$\text{Cost}_{\text{loss}} = C_{\text{E}} \times 91.25 \times \sum_{i=1}^{Ns} \sum_{h=1}^{24} P_{\text{Total\_loss}(i,h)} \tag{3.46}$$

$$\text{Cost}_{\text{Grid}} = C_{\text{Grid}} \times 91.25 \times \sum_{i=1}^{Ns} \sum_{h=1}^{24} P_{\text{Grid}(i,h)} \tag{3.47}$$

$$\text{Cost}_{\text{PV}} = \text{Cost}_{\text{PV}}^{\text{Fixed}} + \text{Cost}_{\text{PV}}^{\text{Variable}} \tag{3.48}$$

$$\text{Cost}_{\text{PV}}^{\text{Fixed}} = \text{CRF} \times C_{\text{PV}} \times P_{\text{sr}} \tag{3.49}$$

$$\text{Cost}_{\text{PV}}^{\text{Variable}} = C_{\text{O\&M}} \times \sum_{i=1}^{Ns} \sum_{h=1}^{24} P_{\text{PV}(i,h)} \tag{3.50}$$

$$\text{CRF} = \frac{\gamma \times (1+\gamma)^{\text{S}}}{(1+\gamma)^{\text{S}} - 1} \tag{3.51}$$

$$\text{Cost}_{\text{DS}} = C_{\text{DS}} \times Q_{\text{DS}} \times \frac{(1+B)^{\text{ND}} \times B}{(1+B)^{\text{ND}} - 1} \tag{3.52}$$

where

$C_{\text{E}}$ is the cost of energy loss in \$/kWh.
$C_{\text{Grid}}$ is the purchasing cost per electric power in \$/kWh.
$C_{\text{DS}}$ is the capital cost of D-STATCOM in \$/kVAR
$C_{\text{O \& M}}$ is the operation & maintenance cost of the PV in \$/kWh.
CRF is the capital recovery factor.
$\gamma$ is the rate of interest on capital investment of the installed PV.
$S$ is the lifetime of PV in years.
$B$ is the asset rate of return.
ND is the lifetime of D-STATCOM in years.
$\text{Cost}_{\text{PV}}^{\text{Fixed}}$ is the fixed cost of the PV unit.
$\text{Cost}_{\text{PV}}^{\text{Variable}}$ is the variable cost of the PV unit.
$P_{\text{Grid}}$ is the imported power from the grid per hour.
$P_{\text{PV}}$ is the generated power by the PV per hour.
$P_{\text{sr}}$ is the rated power of the PV unit.
$Q_{\text{DS}}$ is the rated reactive power of the D-STATCOM in kVAR.

## 3.5 System Constraints

This book determines the optimal allocation of DERs and capacitors in RDS under equality and inequality constraints that can be discussed as follows:

### 3.5.1 *Equality Constraints*

$$P_{\text{Slack}} + \sum_{i=1}^{\text{NPV}} P_{\text{PV},i} = \sum_{i=1}^{\text{NT}} P_{\text{loss},i} + \sum_{i=1}^{\text{NB}} P_{\text{L},i} \tag{3.53}$$

$$Q_{\text{Slack}} + \sum_{i=1}^{\text{NDS}} Q_{\text{DS},i} = \sum_{i=1}^{\text{NT}} Q_{\text{loss},i} + \sum_{i=1}^{\text{NB}} Q_{\text{L},i} \tag{3.54}$$

where $P_{\text{Slack}}$ and $Q_{\text{Slack}}$ represent the substation active and reactive powers, respectively. $P_{\text{L}}$ and $Q_{\text{L}}$ represent the active load demand and reactive load demand, respectively. NPV and NDS denote the number of PV and D-STATCOM units, respectively.

### 3.5.2 *Inequality Constraints*

#### 3.5.2.1 Bus Voltage Constraints

The bus system voltage must be operated between the maximum operating voltage ($V_{max}$) and the minimum operating voltage ($V_{\text{min}}$):

$$V_{\text{min}} \leq V_i \leq V_{\text{max}} \tag{3.55}$$

$$I_n \leq I_{\text{max},n} \quad n = 1, 2, 3 \ldots, \text{NT} \tag{3.56}$$

$$\sum\nolimits_{i=1}^{N_{\text{PV}}} P_{\text{PV}} \leq \rho \times \sum_{i=1}^{\text{NB}} P_{\text{L},i} \tag{3.57}$$

$$\sum_{i=1}^{\text{NDS}} Q_{\text{DST},i} \leq \sum_{i=1}^{\text{NB}} Q_{\text{L},i} \tag{3.58}$$

#### 3.5.2.2 Line Capacity Limits

In this book, the maximum branch currents of the studied RDS are taken into account as follows [139]:

$$I_n \leq I_{\max,n} \quad n = 1, 2, 3 \ldots, \mathrm{NT} \tag{3.59}$$

where $I_{\max,n}$ is the maximum allowable current through branch ($n$).

#### 3.5.2.3 Penetration Level

$$\sum\nolimits_{i=1}^{N_{\mathrm{PV}}} P_{\mathrm{PV}} \leq \rho \times \sum_{i=1}^{\mathrm{NB}} P_{\mathrm{L},i} \tag{3.60}$$

$$\sum_{i=1}^{\mathrm{NDS}} Q_{\mathrm{DST},i} \leq \sum_{i=1}^{\mathrm{NB}} Q_{\mathrm{L},i} \tag{3.61}$$

where $\rho$ denotes the penetration level of the PV unit. min and max are subscripts that indicate the lower and upper allowed levels.

#### 3.5.2.4 Load Demand Modeling

As the load demand is uncertain, normal probability distribution function (pdf) is employed to model it at each node. The normal pdf of the uncertain load demand can be computed as follows [88]:

$$f_n(l) = \frac{1}{\sigma_l \sqrt{2\pi}} \times \exp\left[-\left(\frac{l - \mu_l}{2\sigma_l^2}\right)\right] \tag{3.62}$$

where $f_n(l)$ represents the normal pdf of the load demand; $\mu_l$ and $\sigma_l$ represent the mean and standard deviation of the load demand for each time. The occurrence probability of a segment during a specific hour can be computed as follows:

$$\mathrm{prob}_i^l = \int_{l_i}^{l_{i+1}} f_n(l) \mathrm{d}l \tag{3.63}$$

where $l_i$ and $l_{i+1}$ are the starting and ending points of the interval $i$ and $\mathrm{prob}_i^l$ represents the probability occurrence of interval $i$.

## 3.6 Uncertainty Modeling

The uncertainties in the PV and load demand models are described in this book. In this work, the probabilistic generations of each PV unit and the load demand have been modeled based on the hourly historical data of the location under research. Three years of hourly historical data of load demand and solar irradiance have been considered in this chapter. Based on this, every single year has been split into four seasons. To describe the stochastic behavior of the PV and load demand during each season, a day within that season (24 – h) is considered. Thus, each year has 96 time periods (4 seasons × 24 – h). For each season, the probability distribution function (pdf) of each perod can be obtained by utilizing the data related to the same hours of the day. Accordingly, each period has 270 samples of solar irradiance and load demand (3 years × 3 months per season ×30 days per month) to generate the corresponding hourly pdfs. The probabilistic model of the PV system and load demand can be characterized as follows:

### *3.6.1 Modeling of the Solar Irradiance*

The solar irradiance data of each hour have been used to generate a beta pdf for that hour and can be described as follows [87, 140]:

$$f_{\mathrm{b}}(g_{\mathrm{s}}) = \begin{cases} \dfrac{\Gamma(\alpha+\beta)}{\Gamma(\alpha).\Gamma(\beta)} g_{\mathrm{s}}^{(\alpha-1)}.(1-g_{\mathrm{s}})^{(\beta-1)}, & 0 \le g_{\mathrm{s}} \le 1; \alpha, \beta \ge 0 \\ 0, & \text{Otherwise} \end{cases} \tag{3.64}$$

where $f_{\mathrm{b}}(g_{\mathrm{s}})$ is the beta pdf of the solar irradiance; $\Gamma$ is the gamma function; $\alpha$ and $\beta$ represent the beta parameters for each period. These parameters can be computed by utilizing the historical data as follows [88, 141]:

$$\beta = (1-\mu) \times \left( \frac{\mu \times (1+\mu)}{\sigma^2} - 1 \right) \tag{3.65}$$

$$\alpha = \frac{\mu \times \beta}{1-\mu} \tag{3.66}$$

where $\mu$ and $\sigma$ are the mean and standard deviation of the solar irradiance for each period. The continuous beta pdfs are divided into many segments where each segment generates a mean value and a probability of occurrence. The occurrence probability of a segment during a specific hour can be determined by:

$$\text{prob}_i^{g_s} = \int_{g_{s,i}}^{g_{s,i+1}} f_b(g_s) \mathrm{d}g_{s,i} \tag{3.67}$$

where $g_{s,\ i}$ and $g_{s,\ i+1}$ represent, respectively, the start and end points of the interval $i$. $\text{prob}_i^{g_s}$ represents the probability occurrence of interval i. Based on the generated beta pdf of the solar irradiance of a period, the output power PV for the states of this period can be calculated using (3.32).

### 3.6.2 *Load Demand Modeling*

As the load demand is uncertain, normal pdf is employed to model it at each node. The normal pdf of the uncertain load demand can be computed as follows [142, 143]:

$$f_n(l) = \frac{1}{\sigma_l \sqrt{2\pi}} \times \exp\left[-\left(\frac{l-\mu_l}{2\sigma_l^2}\right)\right] \tag{3.68}$$

where $f_n(l)$ represents the normal pdf of the load demand; $\mu_l$ and $\sigma_l$ represent the mean and standard deviation of the load demand for each time period. The occurrence probability of a segment during a specific hour can be computed as follows:

$$\text{prob}_i^l = \int_{l_i}^{l_{i+1}} f_n(l) \mathrm{d}l \tag{3.69}$$

where $l_i$ and $l_{i+1}$ are the starting and ending points of the interval $i$. $\text{prob}_i^l$ represents the probability occurrence of interval $i$.

## 3.7 Sensitivity Analysis

Sensitivity analysis is applied for all system buses using a forward–backward power flow algorithm and then the buses are arranged in descending order according to their sensitivity values. The buses with high sensitivity values can be defined as a candidate bus for DER and capacitor installation in RDS. The loss sensitivity analysis is used to reduce the search agent of the presented optimization technique and the total simulation time. In this book, loss sensitivity factor (LSF) is utilized to assign the candidate buses for allocation of shunt compensators which prone to more loss reduction with reactive power injection [15]:

$$P_{\text{loss}} = R_{k,k+1}\left(\frac{P_k^2 + Q_k^2}{|V_k|^2}\right) \tag{3.70}$$

Thus, LSF is computed by taking the derivative of the power loss Eq. (3.64) with respect to reactive power injection as shown:

$$\text{LSF} = \frac{\partial P_{\text{loss}}}{\partial Q_k} = R_{k,k+1}\left(\frac{2Q_k}{|V_k|^2}\right) \tag{3.71}$$

LSF is calculated for all nodes and arrange this value in descending order then normalize the bus voltages as follows:

$$V_{\text{norm}} = \frac{|V_k|}{0.95} \tag{3.72}$$

All buses that have $V_{\text{norm}}$ value less than 1.01 and high values of *LSF* are considered the most candidate buses for inclusion of compensators.

**Example 3.1**
Obtain the total active power losses, voltage stability index, voltage deviations, and the objective function for the IEEE-30 bus test system by the backward/forward method. The base voltage and power are 11 kV and 100 MVA, respectively. The line data are in Ω, and the bus data are in kW and kVAR. Bus 1 is taken as the slack bus with its voltage adjusted to 1.05 p.u. (see the MATLAB M-file code in Appendix A)

***Solution*** The calculation of the total active power losses, voltage stability index, and voltage deviations depends on the following values:

Plossb = 805.733
VSIb = 24.9867
VDb = 1.0669
$w1 = 0.5$
$w2 = 0.25$
$w3 = 0.25$

The system with PV and DSTATCOM can be obtained by applying the ALO algorithm (see the MATLAB M-file code in Appendix A). Therefore, the voltage magnitudes in per unit after running the M-file code are as follows:

$$o = (w1 \times \text{TotalActiveLoss\_KW}/\text{Plossb}) + (w2 \times \text{VDD}/\text{VDb}) + (w2 \times 1/\text{Sum\_VSI})$$

$$F_1 = \frac{P_{\text{T,loss}}}{(P_{\text{T,loss}})_{base}} = F_2 = \frac{\text{VD}}{(\text{VD})_{\text{base}}} = F_3 = \frac{1}{\sum_{i=1}^{\text{nb}} |\text{VSI}(i)|_{\text{base}}}$$

The objective function is formulated as follows:

$$F = H_1F_1 + H_2F_2 + H_3F_3 = 0.0980$$

where $H_1$, $H_2$, and $H_3$ are weighting factors. Summation of the weight factors assigned to all impacts must add up to one as:

$$|H_1| + |H_2| + |H_3| = 1$$

The best optimal value of the objective function found by ALO is: 0.098006

$F_1$ represents total active power losses reduction, and it can be found as follows:

$$F_1 = \frac{P_{\text{T,loss}}}{(P_{\text{T,loss}})_{\text{base}}} = \text{TotalActiveLoss\_KW} = 115.9892$$

$F_2$ represents the improving of the voltage profile, which is satisfied by reducing the summation of voltage deviations in RDS and it can be given as follows:

$$F_2 = \frac{\text{VD}}{(\text{VD})_{\text{base}}} = \text{VDD} = 0.0740$$

where $F_3$ represents the voltage stability enhancement which can be achieved by improvement of voltage stability index (*VSI*) as follows:

$$F_3 = \frac{1}{\sum_{i=1}^{\text{nb}} |VSI(i)|_{\text{base}}} =$$

bus_stability_index =

| | |
|---|---|
| 2.0000 | 0.9937 |
| 3.0000 | 1.0027 |
| 4.0000 | 1.0059 |
| 5.0000 | 0.9986 |
| 6.0000 | 0.9936 |
| 7.0000 | 0.9902 |

| | |
|---|---|
| 8.0000 | 0.9894 |
| 9.0000 | 0.9882 |
| 10.0000 | 0.9887 |
| 11.0000 | 0.9882 |
| 12.0000 | 0.9876 |
| 13.0000 | 0.9876 |
| 14.0000 | 0.9879 |
| 15.0000 | 0.9865 |
| 16.0000 | 0.9878 |
| 17.0000 | 0.9889 |
| 18.0000 | 1.0041 |
| 19.0000 | 1.0064 |
| 20.0000 | 1.0133 |
| 21.0000 | 1.0100 |
| 22.0000 | 1.0039 |
| 23.0000 | 0.9993 |
| 24.0000 | 0.9951 |
| 25.0000 | 0.9848 |
| 26.0000 | 0.9839 |
| 27.0000 | 0.9834 |
| 28.0000 | 0.9833 |
| 29.0000 | 0.9823 |
| 30.0000 | 0.9822 |

Sum_VSI = 28.7976

**Example 3.2**
Find the first loss sensitivity index (LSI) for the 10-bus distribution system for case load flow general shown in Fig. 3.4. The base voltage and power are 23 kV and 100 MVA, respectively. The line data are in Ω and the bus data are in kW and kVAR.

***Solution*** The calculation of LSI depends on the values of voltage magnitudes at load buses, total effective active and reactive powers at each load bus, and the resistance of system lines. The values of the voltage magnitudes case load flow general. Therefore, the voltage magnitudes in per unit after running the M-file code are as follows:

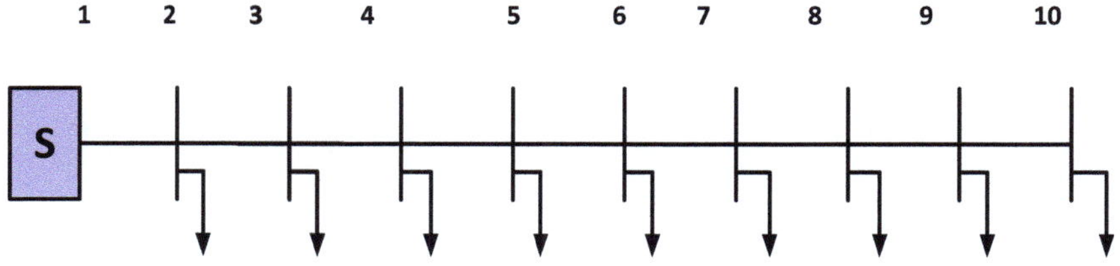


**Fig. 3.4** Single line diagram of the 10-bus radial distribution system

$$V_{\text{mag}} = \begin{bmatrix} V_1 \\ V_2 \\ V_3 \\ V_4 \\ V_5 \\ V_6 \\ V_7 \\ V_8 \\ V_9 \\ V_{10} \end{bmatrix} = \begin{bmatrix} 1.0000 \\ 0.9929 \\ 0.9874 \\ 0.9634 \\ 0.9480 \\ 0.9172 \\ 0.9072 \\ 0.8890 \\ 0.8587 \\ 0.8375 \end{bmatrix} \text{p.u}$$

The total effective active power at each load bus in per unit can be determined as:

$$P_{\text{eff}} = \begin{bmatrix} P_{\text{eff}/2} \\ P_{\text{eff}/3} \\ P_{\text{eff}/4} \\ P_{\text{eff}/5} \\ P_{\text{eff}/6} \\ P_{\text{eff}/7} \\ P_{\text{eff}/8} \\ P_{\text{eff}/9} \\ P_{\text{eff}/10} \end{bmatrix} = \begin{bmatrix} P_2 + P_3 + P_4 + P_5 + P_6 + P_7 + P_8 + P_9 + P_{10} \\ P_3 + P_4 + P_5 + P_6 + P_7 + P_8 + P_9 + P_{10} \\ P_4 + P_5 + P_6 + P_7 + P_8 + P_9 + P_{10} \\ P_5 + P_6 + P_7 + P_8 + P_9 + P_{10} \\ P_6 + P_7 + P_8 + P_9 + P_{10} \\ P_7 + P_8 + P_9 + P_{10} \\ P_8 + P_9 + P_{10} \\ P_9 + P_{10} \\ P_{10} \end{bmatrix} /(1000 \times \text{MV}Ab)$$

$$= \begin{bmatrix} 0.12368 \\ 0.10528 \\ 0.09548 \\ 0.07758 \\ 0.06160 \\ 0.04550 \\ 0.03770 \\ \text{0n26m0} \end{bmatrix}$$

Similarly, the total effective reactive power at each load bus in per unit can be obtained as:

$$Q_{\text{eff}} = \begin{bmatrix} Q_{\text{eff}/2} \\ Q_{\text{eff}/3} \\ Q_{\text{eff}/4} \\ Q_{\text{eff}/5} \\ Q_{\text{eff}/6} \\ Q_{\text{eff}/7} \\ Q_{\text{eff}/8} \\ Q_{\text{eff}/9} \\ Q_{\text{eff}/10} \end{bmatrix} = \begin{bmatrix} Q_2+Q_3+Q_4+Q_5+Q_6+Q_7+Q_8+Q_9+Q_{10} \\ Q_3+Q_4+Q_5+Q_6+Q_7+Q_8+Q_9+Q_{10} \\ Q_4+Q_5+Q_6+Q_7+Q_8+Q_9+Q_{10} \\ Q_5+Q_6+Q_7+Q_8+Q_9+Q_{10} \\ Q_6+Q_7+Q_8+Q_9+Q_{10} \\ Q_7+Q_8+Q_9+Q_{10} \\ Q_8+Q_9+Q_{10} \\ Q_9+Q_{10} \\ Q_{10} \end{bmatrix} /(1000 \times \text{MV}Ab)$$

$$= \begin{bmatrix} 0.04186 \\ 0.03726 \\ 0.03386 \\ 0.02940 \\ 0.01100 \\ 0.00500 \\ 0.00390 \\ 0.00330 \\ 0.00200 \end{bmatrix} \text{p.u}$$

The resistances of system lines in per unit can be calculated as:

$$\text{LSI}_1 = \frac{\partial P_{\text{Loss}k}}{\partial V_q}$$

$$= -2 \times R_k \times \left( \frac{P^2_{(\text{eff}/q)} + Q^2_{(\text{eff}/q)})}{V_q^3} \right)$$

$$R = \begin{bmatrix} R_{1-2} \\ R_{2-3} \\ R_{3-4} \\ R_{4-5} \\ R_{5-6} \\ R_{6-7} \\ R_{7-8} \\ R_{8-9} \\ R_{9-10} \end{bmatrix} \times \frac{\text{MVA}b}{\text{kV}b^2} = \begin{bmatrix} 0.123 \\ 0.014 \\ 0.746 \\ 0.698 \\ 1.983 \\ 0.905 \\ 2.055 \\ 4.795 \\ 5.343 \end{bmatrix} \times \frac{100}{23^2} \text{p.u}$$

$$= \begin{bmatrix} 0.0233 \\ 0.0026 \\ 0.1411 \\ 0.1320 \\ 0.3749 \\ 0.1711 \\ 0.3885 \\ 0.9065 \\ 10101 \end{bmatrix} \text{p.u}$$

Now, the LSI at bus 2 can be calculated as:

$$\begin{aligned} LSI^{\mathrm{Bus2}} &= \frac{\partial P_{\mathrm{Loss}/1-2}}{\partial V_2} \\ &= -2 \times R_{1-2} \times \left( \frac{P^2_{(\mathrm{eff}/2)} + Q^2_{(\mathrm{eff}/2)})}{V_2^3} \right) \\ &= -2 \times 0.0233 \times \left( \frac{0.12368^2 + 0.04186^2}{0.9929^3} \right) \\ &= -0.00081165 \text{ p.u.} \end{aligned}$$

Similarly, the LSI1 for the rest of loads buses are:

$$\begin{array}{llll} \mathrm{LSI}_1^{\mathrm{Bus3}} = -6.8581 \times 10^{-5} & \text{p.u.} \quad , & LSI_1^{\mathrm{Bus\ 4}} = -0.0032384 & \text{p.u.} \\ \mathrm{LSI}_1^{\mathrm{Bus\ 5}} = -0.00213310 & \text{p.u.} \quad , & LSI_1^{\mathrm{Bus\ 6}} = -0.0038051 & \text{p.u} \\ \mathrm{LSI}_1^{\mathrm{Bus\ 7}} = -0.00096061 & \text{p.u.} \quad , & LSI_1^{\mathrm{Bus\ 8}} = -0.0015889 & \text{p.u} \\ \mathrm{LSI}_1^{\mathrm{Bus\ 9}} = -0.00199670 & \text{p.u.} \quad , & LSI_1^{\mathrm{Bus\ 10}} = -0.00093873 & \text{p.u} \end{array}$$

**Example 3.3**

Obtain the total active power losses, voltage stability index, voltage deviations for the 10-bus distribution system with PV and DSTATCOM shown in Fig. 3.4. The base voltage and power are 23 kV and 100 MVA, respectively. The line data are in Ω and the bus data are in kW and kVAR.

***Solution*** The calculation of the total active power losses, voltage stability index, and voltage deviations depends on the following values:

Plossb = 783.669
VSIb = 6.6131
VDb = 0.6988
$w1 = 0.5$
$w2 = 0.25$
$w3 = 0.25$

The system with PV and DSTATCOM can be obtained by applying the ALO algorithm. The system with PV and DSTATCOM can be obtained by applying the ALO algorithm as following:

$F_1$ represents total active power losses reduction and it can be found as follows:

$F_1 = \frac{P_{\mathrm{T,loss}}}{(P_{\mathrm{T,loss}})_{\mathrm{base}}} =$

=====================================

bus no. mag_V angle_V Pg_KW Qg_KVAR

=====================================

BusOut =

1.0e+04 *

| | | | | |
|---|---|---|---|---|
| 0.0001 | 0.0001 | 0 | 0 | 0 |
| 0.0002 | 0.0001 | −0.0000 | 0 | 0 |
| 0.0003 | 0.0001 | 0.0000 | 0 | 0 |
| 0.0004 | 0.0001 | 0.0000 | 0 | 0 |
| 0.0005 | 0.0001 | 0.0001 | 0 | 0 |
| 0.0006 | 0.0001 | 0.0001 | 1.2368 | 0 |
| 0.0007 | 0.0001 | 0.0000 | 0 | 0.4186 |
| 0.0008 | 0.0001 | −0.0000 | 0 | 0 |
| 0.0009 | 0.0001 | −0.0001 | 0 | 0 |
| 0.0010 | 0.0001 | −0.0001 | 0 | 0 |

=====================================

Line no. from to mag_I(p.u.) Ploss(KW) Qloss(KVAR)

=====================================

LineFlow =

| | | | | | |
|---|---|---|---|---|---|
| 1.0000 | 1.0000 | 2.0000 | 0.0498 | 0.0578 | 0.1936 |
| 2.0000 | 2.0000 | 3.0000 | 0.1456 | 0.0561 | 2.4259 |
| 3.0000 | 3.0000 | 4.0000 | 0.2478 | 8.6596 | 13.9821 |
| 4.0000 | 4.0000 | 5.0000 | 0.4313 | 24.5628 | 21.3975 |
| 5.0000 | 5.0000 | 6.0000 | 0.6420 | 154.5320 | 134.6223 |
| 6.0000 | 6.0000 | 7.0000 | 0.5697 | 55.5445 | 48.3844 |
| 7.0000 | 7.0000 | 8.0000 | 0.3799 | 56.0771 | 31.7603 |
| 8.0000 | 8.0000 | 9.0000 | 0.2677 | 64.9770 | 36.8022 |
| 9.0000 | 9.0000 | 10.0000 | 0.1686 | 28.7298 | 16.2720 |

=====================================
===========================

Minimum voltage (p.u.): 0.97963 @ bus 10
Maximum voltage (p.u.) excluding the slack bus: 1.03928 @ bus 7
Total active load demand (KW): 12368.000
Total reactive load demand (KVAR): 4186.000
Total active loss (KW): 393.197
Total reactive loss (KVAR): 305.840

$F_2$ represents the improving of the voltage profile, which is satisfied by reducing the summation of voltage deviations in RDS and it can be given as follows:

$$F_2 = \frac{\mathrm{VD}}{(\mathrm{VD})_{\mathrm{base}}} =$$

0.0393 − 0.0306i
−0.1447 + 0.0155i
−0.2428 + 0.0493i
−0.4213 + 0.0926i
−0.5811 + 0.2731i
0.4502 + 0.3491i
0.3774–0.0435i
0.2651–0.0376i
0.1670–0.0235i

$$\mathrm{VDD} = 0.1435$$

where $F_3$ represents the voltage stability enhancement which can be achieved by improvement of voltage stability index (*VSI*) as follows:

$$F_3 = \frac{1}{\sum_{i=1}^{nb} |\mathrm{VSI}(i)|_{\mathrm{base}}} =$$

bus_stability_index =

| | |
|---|---|
| 2.0000 | 0.9987 |
| 3.0000 | 0.9995 |
| 4.0000 | 1.0177 |
| 5.0000 | 1.0449 |
| 6.0000 | 1.1757 |
| 7.0000 | 1.1666 |
| 8.0000 | 1.0973 |
| 9.0000 | 0.9890 |
| 10.0000 | 0.9197 |

VSI_min =

0.9197

Sum_VSI =

9.4092

**Example 3.4**
Find the best optimal value of the objective function by ALO for the 12-bus distribution shown in Figs. 3.5 and 3.6 system with PV and DSTATCOM and obtain optimal sizing and location. The base voltage and power are 11 kV and 100 MVA, respectively. The line data are in Ω and the bus data are in kW and kVAR.

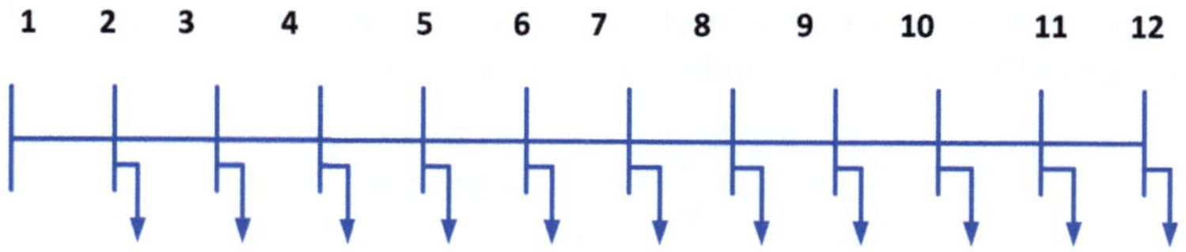


**Fig. 3.5** Single line diagram of the 12-bus radial distribution system

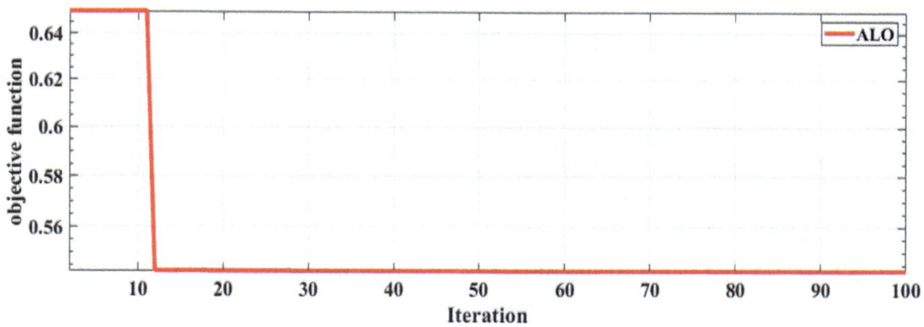


**Fig. 3.6** The convergence characteristic of the ALO objective function

***Solution*** The calculation of the objective function and obtainment of optimal sizing and location depend on the following values:

Plossb = 20.7138
VSIb = 9.4954
VDb = 0.4020
$w1 = 0.5$
$w2 = 0.25$
$w3 = 0.25$

**Objective Function**

The objective function is formulated as follows:

$$F = H_1F_1 + H_2F_2 + H_3F_3$$

where $H_1$, $H_2$, and $H_3$ are weighting factors. Summation of the weight factors assigned to all impacts must add up to one as:

$$|H_1| + |H_2| + |H_3| = 1$$

$F_1$ is the first objective of the multi objective function which represents total active power losses reduction, and it can be found as follows:

$$F_1 = \frac{P_{\mathrm{T,loss}}}{(P_{\mathrm{T,loss}})_{\mathrm{base}}} = 20.7138$$

$F_2$ represents the improving of the voltage profile, which is satisfied by reducing the summation of voltage deviations in RDS and it can be given as follows:

$$F_2 = \frac{\mathrm{VD}}{(\mathrm{VD})_{\mathrm{base}}} = 0.4020$$

where $F_3$ represents the voltage stability enhancement which can be achieved by improvement of voltage stability index (*VSI*) as follows:

$$F_3 = \frac{1}{\sum_{i=1}^{nb} |\mathrm{VSI}(i)|_{\mathrm{base}}} = 9.4954$$

The best optimal value of the objective function found by ALO is: 0.54249

======================================

Vbus =

1.0000 + 0.0000i
0.9999–0.0000i
1.0005–0.0004i
0.9922 + 0.0027i
0.9815 + 0.0064i
0.9783 + 0.0076i
0.9755 + 0.0086i
0.9671 + 0.0126i
0.9592 + 0.0163i
0.9563 + 0.0175i
0.9555 + 0.0179i
0.9552 + 0.0180i

IB =

0.0001 − 0.0001i
−0.0005 + 0.0005i
0.0035–0.0032i
0.0029–0.0027i
0.0026–0.0023i
0.0024–0.0022i
0.0019–0.0016i

```
 0.0014–0.0012i
 0.0010–0.0008i
 0.0006–0.0005i
 0.0002–0.0002i

   VDD =

0.2790

   VDD2 =

0.0101

   V_mag =

1.000000000000000
0.999851179490182
1.000522169547643
0.992174014649502
0.981552809884612
0.978307002323343
0.975554937949232
0.967221279945066
0.959289963540436
0.956509763408107
0.955622909823004
0.95541672750196

   MinVoltage =

0.9554

   MaxVoltage =

1

   bus_stability_index =

 2.0000   0.9994
 3.0000   1.0021
 4.0000   0.9689
 5.0000   0.9279
 6.0000   0.9160
 7.0000   0.9057
 8.0000   0.8748
 9.0000   0.8465
```

```
10.0000   0.8370
11.0000   0.8339
12.0000   0.8332

  VSI_min =

0.8332

  Sum_VSI =

9.9454

  Q_Loss_sensetivity_index =

 3.0000   -0.0011
 1.0000    0
 2.0000    0.0001
12.0000    0.0003
11.0000    0.0012
 7.0000    0.0037
10.0000    0.0038
 6.0000    0.0043
 4.0000    0.0112
 9.0000    0.0114
 8.0000    0.0123
 5.0000    0.0142

  Voltage_Loss_sensetivity_index =

 5.000000000000000   -0.000083889212794
 4.000000000000000   -0.000078289523576
 8.000000000000000   -0.000045921717473
 9.000000000000000   -0.000031972916046
 6.000000000000000   -0.000022897234630
 7.000000000000000   -0.000018119035210
10.000000000000000   -0.000007500821015
11.000000000000000   -0.000001448307598
 3.000000000000000   -0.000000993839255
12.000000000000000   -0.000000105584233
2.000000000000000   -0.000000041850221
1.000000000000000    0
```

```
==========================================
bus no.   mag_V   angle_V   Pg_KW   Qg_KVAR
==========================================
BusOut =

 1.0000   1.0000    0    0    0
 2.0000   0.9999   -0.0003    0    0
 3.0000   1.0005   -0.0202   435.0000   405.0000
 4.0000   0.9922   0.1545    0    0
 5.0000   0.9816   0.3757    0    0
 6.0000   0.9783   0.4432    0    0
 7.0000   0.9756   0.5023    0    0
 8.0000   0.9672   0.7491    0    0
 9.0000   0.9593   0.9743    0    0
10.0000   0.9565   1.0482    0    0
11.0000   0.9556   1.0713    0    0
12.0000   0.9554   1.0782    0    0

==========================================
Line no.   from to mag_I(p.u.) Ploss(KW) Qloss(KVAR)
==========================================
LineFlow =

 1.0000    1.0000    2.0000   0.0002   0.0021    0.0009
 2.0000    2.0000 3.0000   0.0007   0.0497    0.0207
 3.0000    3.0000    4.0000   0.0047   3.8838    1.6184
 4.0000    4.0000    5.0000 0.0040   4.1171    1.7163
 5.0000    5.0000    6.0000   0.0035   1.1200    0.4663
 6.0000    6.0000    7.0000   0.0033   0.8838    0.3678
 7.0000    7.0000    8.0000   0.0025   2.2208    0.6128
 8.0000    8.0000    9.0000   0.0018   1.5336    0.4341
 9.0000    9.0000   10.0000   0.0012   0.3587    0.1015
10.0000   10.0000   11.0000   0.0007   0.0692    0.0196
11.0000   11.0000   12.0000   0.0002   0.0050    0.0014

==========================================
Minimum voltage (p.u.): 0.95542 @ bus 12
Maximum voltage (p.u.) excluding the slack bus: 1.00000 @ bus 1
Total active load demand (KW): 435.000
Total reactive load demand (KVAR): 405.000
Total Active Loss (KW): 14.244
```

Total Reactive Loss (KVAR): 5.360
Total energy losses cost ($): 7486.611

==========================================

$o =$

0.5425

Location_PV1 =

3

Location_PV2 =

3

Location_DSTATCOM1 =

3

Location_DSTATCOM2 =

3

Size_PV1 =

435

Size_PV2 =

435

Size_DSTATCOM1 =

405

Size_DSTATCOM2

405

**Example 3.5**
Find the second loss sensitivity index (LSI2) as in for the 10-bus distribution system. The base voltage and power are 23 kV and 100 MVA, respectively. The line data are in Ω and the bus data are in kW and kVAR (Fig. 3.7).

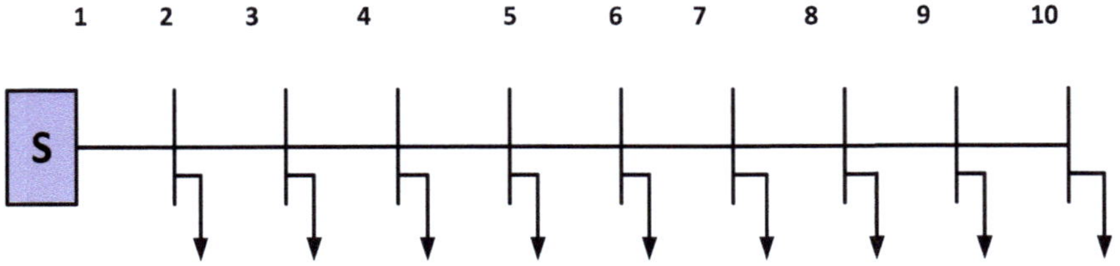


**Fig. 3.7** Single line diagram of the 10-bus radial distribution system

***Solution*** The calculation of LSI2 depends on the values of voltage magnitudes at load buses, total effective reactive power at each load bus, and the resistance of system.

$$V_{\text{mag}} = \begin{bmatrix} V_1 \\ V_2 \\ V_3 \\ V_4 \\ V_5 \\ V_6 \\ V_7 \\ V_8 \\ V_9 \\ V_{10} \end{bmatrix} = \begin{bmatrix} 1.0000 \\ 0.9929 \\ 0.9874 \\ 0.9634 \\ 0.9480 \\ 0.9172 \\ 0.9072 \\ 0.8890 \\ 0.8587 \\ 0.8375 \end{bmatrix} \text{ p.u.}$$

The total effective active power at each load bus in per unit can be determined as:

$$P_{\text{eff}} = \left[ P_{\frac{\text{eff}}{2}} P_{\frac{\text{eff}}{3}} P_{\frac{\text{eff}}{4}} P_{\frac{\text{eff}}{5}} P_{\frac{\text{eff}}{6}} P_{\frac{\text{eff}}{7}} P_{\frac{\text{eff}}{8}} P_{\frac{\text{eff}}{9}} P_{\frac{\text{eff}}{10}} \right]$$

$$= \begin{bmatrix} P_2 + P_3 + P_4 + P_5 + P_6 + P_7 + P_8 + P_9 + P_{10} \\ P_3 + P_4 + P_5 + P_6 + P_7 + P_8 + P_9 + P_{10} \\ P_4 + P_5 + P_6 + P_7 + P_8 + P_9 + P_{10} \\ P_5 + P_6 + P_7 + P_8 + P_9 + P_{10} \\ P_6 + P_7 + P_8 + P_9 + P_{10} \\ P_7 + P_8 + P_9 + P_{10} \\ P_8 + P_9 + P_{10} \\ P_9 + P_{10} \\ P_{10} \end{bmatrix} =$$

$$= \begin{bmatrix} 1840 + 980 + 1790 + 1598 + 1610 + 780 + 1150 + 980 + 1640 \\ 980 + 1790 + 1598 + 1610 + 780 + 1150 + 980 + 1640 \\ 1790 + 1598 + 1610 + 780 + 1150 + 980 + 1640 \\ 1598 + 1610 + 780 + 1150 + 980 + 1640 \\ 1610 + 780 + 1150 + 980 + 1640 \\ 780 + 1150 + 980 + 1640 \\ 1150 + 980 + 1640 \\ 980 + 1640 \\ 1640 \end{bmatrix} /10^5$$

$$= \begin{bmatrix} 0.12368 \\ 0.10528 \\ 0.09548 \\ 0.07758 \\ 0.06160 \\ 0.04550 \\ 0.03770 \\ 0.02620 \\ 0.01640 \end{bmatrix} \text{p.u}$$

Similarly, the total effective reactive power at each load bus in per unit can be obtained as:

$$Q_{\text{eff}} = \begin{bmatrix} Q_{\text{eff}/2} \\ Q_{\text{eff}/3} \\ Q_{\text{eff}/4} \\ Q_{\text{eff}/5} \\ Q_{\text{eff}/6} \\ Q_{\text{eff}/7} \\ Q_{\text{eff}/8} \\ Q_{\text{eff}/9} \\ Q_{\text{eff}/10} \end{bmatrix}$$

$$= \begin{bmatrix} Q_2 + Q_3 + Q_4 + Q_5 + Q_6 + Q_7 + Q_8 + Q_9 + Q_{10} \\ Q_3 + Q_4 + Q_5 + Q_6 + Q_7 + Q_8 + Q_9 + Q_{10} \\ Q_4 + Q_5 + Q_6 + Q_7 + Q_8 + Q_9 + Q_{10} \\ Q_5 + Q_6 + Q_7 + Q_8 + Q_9 + Q_{10} \\ Q_6 + Q_7 + Q_8 + Q_9 + Q_{10} \\ Q_7 + Q_8 + Q_9 + Q_{10} \\ Q_8 + Q_9 + Q_{10} \\ Q_9 + Q_{10} \\ Q_{10} \end{bmatrix} /(1000 \times \text{MVA}b)$$

$$= \begin{bmatrix} 0.04186 \\ 0.03726 \\ 0.03386 \\ 0.02940 \\ 0.01100 \\ 0.00500 \\ 0.00390 \\ 0.00330 \\ 0.00200 \end{bmatrix} \text{p.u}$$

The resistances of system lines in per unit can be calculated as:

$$R = \begin{bmatrix} R_{1-2} \\ R_{2-3} \\ R_{3-4} \\ R_{4-5} \\ R_{5-6} \\ R_{6-7} \\ R_{7-8} \\ R_{8-9} \\ R_{9-10} \end{bmatrix} \times \frac{\mathrm{MV}Ab}{\mathrm{kV}b^2} = \begin{bmatrix} 0.123 \\ 0.014 \\ 0.746 \\ 0.698 \\ 1.983 \\ 0.905 \\ 2.055 \\ 4.795 \\ 5.343 \end{bmatrix} \times \frac{100}{23^2}\mathrm{p.u}$$

$$= \begin{bmatrix} 0.0233 \\ 0.0026 \\ 0.1411 \\ 0.1320 \\ 0.3749 \\ 0.1711 \\ 0.3885 \\ 0.9065 \\ 1.0101 \end{bmatrix} \mathrm{p.u}$$

The calculation of LSI2 depends on the values of voltage magnitudes at load buses, total effective reactive power at each load bus and the resistance of system lines. Therefore, the LSI2 at bus 2 can be calculated as:

$$\begin{aligned} \mathrm{LSI}_1^{\mathrm{Bus}}2 &= \frac{\partial P_{\mathrm{Loss}/1-2}}{\partial Q_{\mathrm{eff}/2}} \\ &= \frac{2 \times Q_{\mathrm{eff}/2 \times R_{1-2}}}{V_2^2} \\ &= \frac{2 \times 0.04186 \times 0.0233}{0.9929^2} \\ &= 0.001979 \text{ p.u.} \end{aligned}$$

# Chapter 4
# Optimal Allocation of Distributed Energy Resources Using Modern Optimization Techniques

## 4.1 Metaheuristic Optimization Techniques

### *4.1.1 Marine Predators Algorithm (MPA)*

The marine predators algorithm (MPA) is a modern technique that mimics the marine predators foraging such as tunas, marlines, swordfish, sunfish, and sharks with their motion in oceans [144]. The marine predators foraging phase is based on two motions, comprising the Brownian motion and the Lévy flight walk motion, shown in Fig. 4.1. It is worth mentioning that the Lévy flight walk is an arbitrary process of a transition of an object from one place to another using a probability distribution factor [23].

The predators move in a Brownian movement pattern when these predators exist in a region that has prey abundant. In contrast, the predators move like Lévy flight walk at the food in a prey-sparse environment. Besides that, the predators in sudden vertical jumps are recorded by the Fish Aggregating Device (FAD). The steps procedure and the mathematical representation of the MPA are discussed in the following:

### *4.1.2 Initialization*

The initial populations are generated randomly using (4.1) as follows:

$$X_i = X_i^{\min} + \left(X_i^{\max} - X_i^{\min}\right) \times \text{rand} \tag{4.1}$$

where $X_i^{\max}$ means the maximum limit of variable *I*, while $X_i^{\min}$ denotes to its minimum limit.

rand∈ [0–1]. Then, the fitness function is calculated as follows:



A. M. A. El-Hameid et al., *Enhancement of Grid-Connected Photovoltaic Systems Using Artificial Intelligence*, https://doi.org/10.1007/978-3-031-29692-5_4

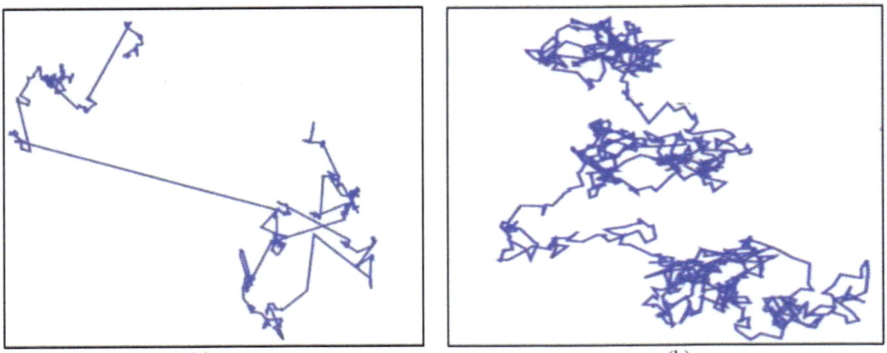


**Fig. 4.1** Trajectories (**a**) Lévy flight motion and (**b**) Brownian motion

$$F_i = \text{obj}\ (X_i) \tag{4.2}$$

### *4.1.3 Assigning the Top Predator*

A matrix is constructed to organize the populations which are called the prey matrix:

$$X = \begin{bmatrix} X_{1,1} & X_{1,2} & \cdots & X_{1d} \\ X_{2,1} & X_{22} & \cdots & X_{2,1} \\ \vdots & \vdots & \ddots & \vdots \\ X_{n,1} & X_{n,2} & \ldots & X_{n,d} \end{bmatrix} \tag{4.3}$$

where $n$ denotes the population number and $d$ is the number of control variables. The top predators are assigned by arrangement the populations based on their fitness functions. The elite matrix is constructed, which represents the top predators:

$$E = \begin{bmatrix} E_{1,1} & E_{1,2} & \cdots & E_{1d} \\ E_{2,1} & E_{22} & \cdots & E_{2,1} \\ \vdots & \vdots & \ddots & \vdots \\ E_{n,1} & E_{n,2} & \ldots & E_{n,d} \end{bmatrix} \tag{4.4}$$

### *4.1.4 The Brownian and Lévy Flight Orientations*

The locations of the predators and the prey are updated using three phases, which are based on the ratio of the velocity between the predator and the prey.

Phase 1: This phase is implemented when the predator's velocity is more than the prey's velocity. The phase mimics the exploration process of the algorithm, and the predator and the prey move like a Brownian movement pattern, which can be formulated as follows:

$$\overline{\mathrm{SEZ}_i} = \overline{R_{\mathrm{Br}}} \bigoplus \left(\overline{E_i} - \overline{R_{\mathrm{Br}}} \bigoplus \overline{X_i}\right) \quad \text{if } T \leq T_{\max} \tag{4.5}$$

$$\overline{X_i} = \overline{X_i} + P.\bar{R} \bigoplus \overline{\mathrm{SEZ}_i} \tag{4.6}$$

where $T$ is the current iteration, while $T_{\max}$ represents the maximum number of iterations, respectively; $\bigoplus$ represents the entry wise multiplication. $\overline{R_{\mathrm{Br}}}$ denotes a vector that includes random numbers from the Brownian motion. $\overline{\mathrm{SEZ}_i}$ is a step size vector, while $P$ is a constant number, which is equal to 0.5.

Phase 2: This phase is implemented when the velocities of the predator and the prey are identical. Thus, the book is considered an intermitted phase between the exploitation and the exploration processes. The populations are separated into two sections: the first section is utilized for the exploration, while the second section is applied for the exploitation. Thus, this step can be represented as follows:

$$\text{if } \frac{1}{3} T_{\max} \leq T \leq \frac{2}{3} T_{\max}$$

$$\overline{\mathrm{SEZ}_i} = \overline{R_{\mathrm{Levy}}} \bigoplus \left(\overline{E_i} - \overline{R_{\mathrm{Levy}}} \bigoplus \overline{X_i}\right) \quad \text{for } i = 1,2,3 \ldots, \quad \frac{n}{2} \tag{4.7}$$

$$\overline{X_i} = \overline{X_i} + P.\bar{R} \bigoplus \overline{\mathrm{SEZ}_i} \tag{4.8}$$

The second section is represented as follows:

$$\overline{\mathrm{SEZ}_i} = \overline{R_{\mathrm{Br}}} \bigoplus \left(\overline{E_i} - \overline{R_{\mathrm{Br}}} \bigoplus \overline{X_i}\right) \qquad \text{for } i = \frac{n}{2}, \ldots, n \tag{4.9}$$

$$\overline{X_i} = \overline{X_i} + P.\mathrm{CF} \bigoplus \overline{\mathrm{SZ}_i} \tag{4.10}$$

where $R_{\mathrm{Levy}}$ denotes a vector that mimics the motion of prey in Lévy flight motion. CF represents the adaptive operator which is used for controlling the step size of predator motion.

Phase 3: This phase represents the exploitation process used when the predator's velocity is more than the prey's velocity. In this stage, the populations or the predator moves in Lévy flight orientation, which can be represented as follows:

$$\overline{\mathrm{SEZ}_i} = \overline{R_{\mathrm{Levy}}} \bigoplus \left(\overline{R_{\mathrm{Levy}}} \bigoplus \overline{E_i} - \overline{X_i}\right) \quad \text{for } i = 1,2,3 \ldots, \quad \frac{n}{2} \tag{4.11}$$

$$\overline{X_i} = \overline{E_i} + P.\bar{R} \bigoplus \overline{\mathrm{SEZ}_i} \tag{4.12}$$

### 4.1.5 FDAs Effect and Eddy Formation

Most sharks spent about 80% of their time in the closed region, while 20% of their time moves to new areas are known as Fish Aggregating Devices (FADs) or the eddy formation. Representation of this step is formulated as follows:

$$\overline{X_i} = \begin{cases} \overline{X_i} + \mathrm{CF}\left[X_i^{\min} + R\left(X_i^{\max} - X_i^{\min}\right)\right] \oplus \bar{U} & \text{If } r \leq \text{FADS} \\ \overline{X_i} + \left[\text{FADS}(1-r) + r\right]\left(\overline{X_{r1}} - \overline{X_{r2}}\right) & \text{If } r > \text{FADS} \end{cases} \tag{4.13}$$

where $r$ denotes a random value $r \in [0\text{–}1]$. $r1$ and $r2$ are indices selected from the prey matrix. FADs are the probability of the FADs which is 0.2. $\bar{U}$ represents a binary vector.

### 4.1.6 Marine Memory

The marine predators remember the optimal location efficiently for foraging. Therefore, in the MPA, the new solution is compared to those in the last iteration to obtain the best solution. Figure 4.2 shows the flowchart of application the MPA for optimal photovoltaic distributed generation (PV-DG) and distributed static compensator (D-STATCOM).

## 4.2 Equilibrium Optimizer (EO)

EO is a recent physical-based algorithm that mimics the mass balance of a dynamic mixture on a control volume [145]. The mass balance equation consists of input, output, and generated mass in a control volume and can be given as follows:

$$V\frac{\mathrm{d}c}{\mathrm{d}t} = QC_{\mathrm{eq}} - QC + G \tag{4.14}$$

where $V$ represents the control volume, $C$ is the concentration of $V$, $Q$ is the input and output flow rate, $Ceq$ represents the equilibrium state concentration, and $G$ represents the mass generation rate. By integrating and manipulating Eq. (4.14) the following is obtained:

$$C = C_{\mathrm{eq}} + \left(C_0 - C_{\mathrm{eq}}\right)\exp[-\lambda(t - t_0)] + \frac{G}{\lambda V}\left(1 - \left(\exp[-\lambda(t - t_0)]\right)\right) \tag{4.15}$$

where $\lambda = \left(\frac{Q}{V}\right)$. $C_0$ represents the initial concentration while $t_0$ denotes the initial start time. Equation (4.15) is the main updating equation of the EO where the

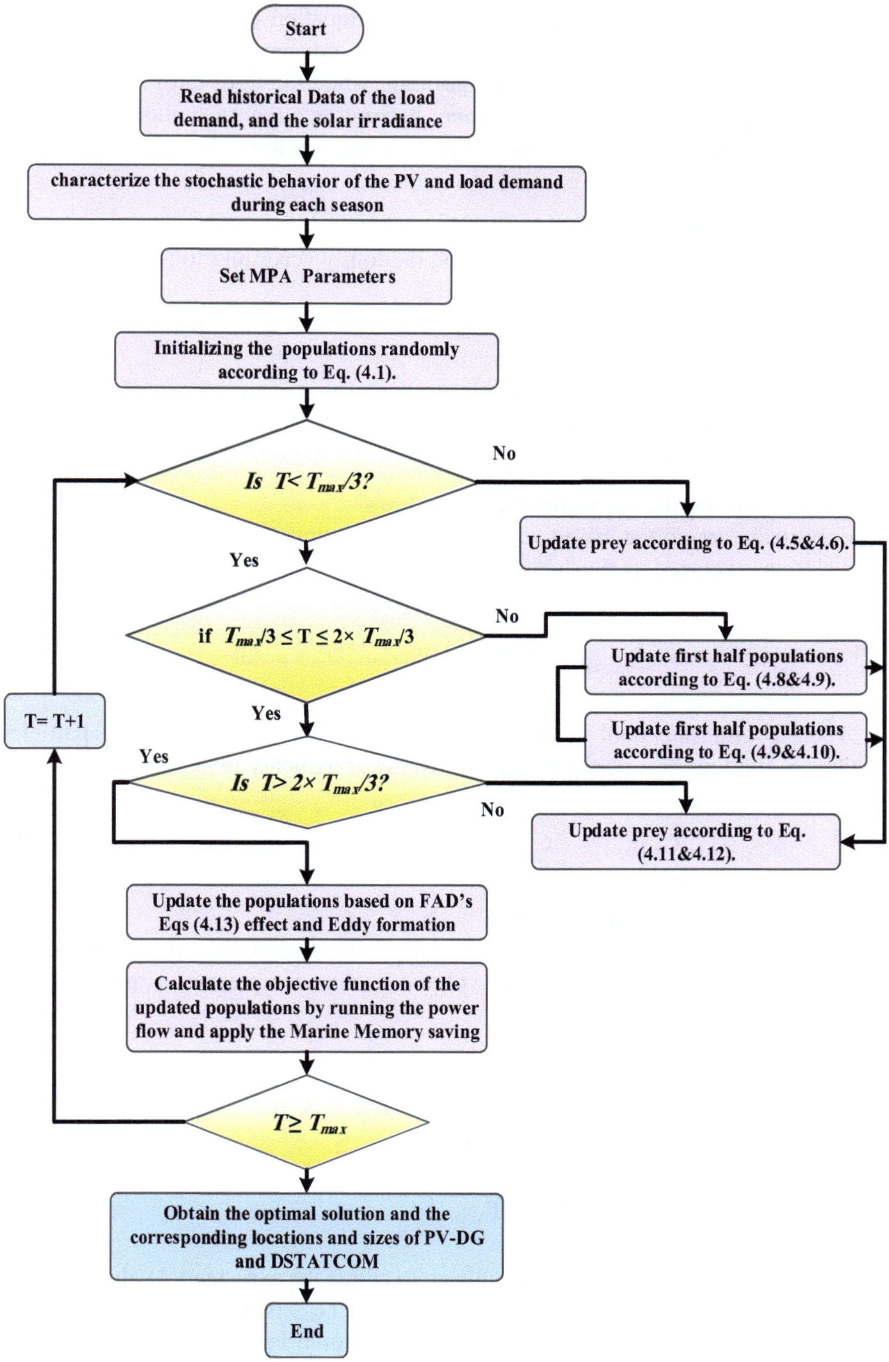


**Fig. 4.2** The flowchart of application the MPA for optimal PV-DG and D-STATCOM

concentration represents the particle's position. Equation (4.15) depends on three terms. The first one is related to the equilibrium concentration, the second term denotes the exploration phase of the algorithm, while the third term represents the exploitation phase of the algorithm where it represents the generation rate, which enables the solution to move in small steps. The steps of the EO can be summarized as follows:

**Step 1: Initialization**
Obtain the initial particles (concentrations) randomly according to (4.16)

$$C_i^{\text{initial}} = C_{\min} + \text{rand}_i \left(C_{\max} - C_{\min}\right) \qquad i = 1, 2, \ldots\ldots n \tag{4.16}$$

where $C_{\min}$ and $C_{\max}$ represent the minimum and maximum values of the control variables. $\text{rand}_i$ is the random variable within [0,1]. Then, calculate the objective function for the initial points.

**Step 2: Assigning the Equilibrium Candidates**
The particles are sorted based on their corresponding values, and the best four particles are averaged as illustrated in (4.17).

$$\vec{C}_{\text{eq(avg)}} = \frac{\vec{C}_{\text{eq1}} + \vec{C}_{\text{eq2}} + \vec{C}_{\text{eq3}} + \vec{C}_{\text{eq4}}}{4} \tag{4.17}$$

Then, equilibrium pool vector is constructed which includes the five particles which are the four best particles, and the averaged particles are as follows:

$$\vec{C}_{\text{eq,pool}} = \left\{\vec{C}_{\text{eq1}}, \vec{C}_{\text{eq2}}, \vec{C}_{\text{eq3}}, \vec{C}_{\text{eq4}}, \vec{C}_{\text{eq(avg)}}\right\} \tag{4.18}$$

**Step 3: Updating the concentration**
The concentrations are updated exponentially with iteration based on exponential term ($F$) which is based on the control volume $\lambda$, where it represents a random vector within the range [0, 1]:

$$\vec{F} = a_1 \text{sign}\left(\vec{r} - 0.5\right)\left[e^{-\vec{\lambda} t} - 1\right] \tag{4.19}$$

$$t = \left(1 - \frac{T}{T_{\text{Max}}}\right)^{\left(a_2 \frac{T}{T_{\text{Max}}}\right)} \tag{4.20}$$

where $a_1$ and $a_2$ are constant values utilized to adjust the exponential value and they are selected to be 2 and 1, respectively, and $T$ and $T_{\text{Max}}$ present the current and the maximum number of iterations, respectively. It should be highlighted here that $a_1$ controls the exploration quantity of the algorithm while $a2$ controls the exploitation feature of the algorithm. Sign ($r$ – 0.5) controls the exploration direction.

**Step 4: Update of the concentration depends upon the generation rate.**
The populations are updated in this step by applying the generation rate method for enhancing the exploitation process of this algorithm which can be described as follows:

$$\vec{G} = \overrightarrow{G_0}\ e^{-\vec{k}\,(t - t_0)} \tag{4.21}$$

$$\overrightarrow{G_0} = \overrightarrow{\mathrm{GCP}}\ \left(\overrightarrow{C_{\mathrm{eq}}} - \vec{\lambda}\,\vec{C}\right) \tag{4.22}$$

$$\overrightarrow{\mathrm{GCP}} = \begin{cases} 0.5\, r_1 & r_2 \geq GP \\ 0 & r_2 < GP \end{cases} \tag{4.23}$$

where $r_1$ and $r_2$ are random numbers in [0, 1] and GCP vector is constructed by the repetition of the same value resulted from. GP is the generation probability that controls the participation probability of concentration updating by the generation rate. GP = 1 means that there will be no generation rate term participating in the optimization process. GP = 0 means that the generation rate term will be participating in the process. GP = 0.5 provides a good balance between exploration and exploitation phases.

The updated equation of EO based on the previous steps can be formulated as follows:

$$\vec{C} = \overrightarrow{C_{\mathrm{eq}}} + \left(\vec{C} - \overrightarrow{C_{\mathrm{eq}}}\right).\vec{F} + \frac{\vec{G}}{\vec{\lambda}\,V}\left(1 - \vec{F}\right) \tag{4.24}$$

**Step 5: Adding memory saving.**
In this step, the concentration will track the obtained best coordination where the objective function of each particle is compared with its value in the previous iteration. A flowchart of the EO with sensitivity analysis for allocation of PV and D-STATCOM is illustrated in Fig. 4.3.

The procedure of application the EO for optimal allocation the PV unit and D-STATCOM can summarized as follows.

Step 1: Set the parameters of the EO and the system data including the line data and the bus data.
Step 2: Run the load flow and carry out the sensitivity analysis of the system and assign the high potential buses for allocation the PV unit and D-STATCOM.
Step 3: Initialize the populations randomly.
Step 4: Calculate the objective function of the obtained populations by running the load flow.
Step 5: Determine the equilibrium candidates (the best four solution and their average values).
Step 6: Construct F, G0, and GCP according to (4.19), (4.22), and (4.23), respectively.

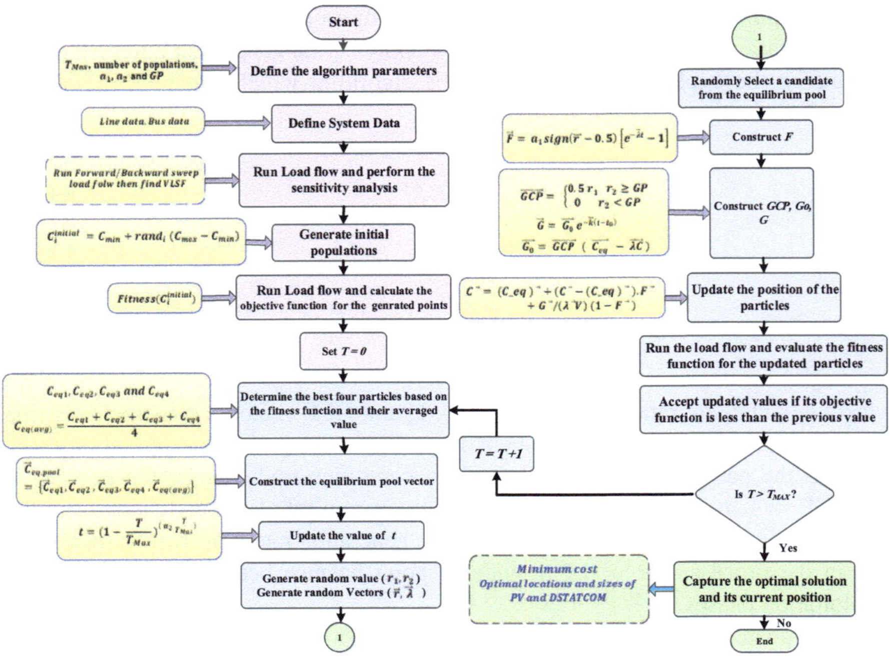


**Fig. 4.3** Solution process of EO and sensitivity analysis for allocation of PV and D-STATCOM

Step 7: Update the locations of the population according to (4.24).

Step 8: Accept the new updated population its fitness function is better that the previous solution.

Step 9: Go to step (4) if the stopping criteria is not stratified. Otherwise, end the program and obtain the best solution (the optimal locations and ratings of the PV and D-STATCOM and the corresponding fitness function).

## 4.3 Lightning Attachment Procedure Optimization (LAPO)

LAPO is a novel physical-based algorithm that mimics the formation of lightning in nature. Lightning happens when a large amount of the positive and negative charges accumulate in the cloud. When the amount of these charges increases, the electric voltage will increase leading to air breakdown. Consequently, the lightning occurs in several paths including the downward and upward leaders. Figure 4.4 depicts the formation of the positive and negative charges inside the cloud as well as the formation of the downward leader and the upward leaders. Referring to this figure, the lightning starts from several points which are known as the initial spots of the leaders. Also, it can be obvious that the direction of the upper ward leaders and the downward leaders are in opposite direction, this sequence leads to formation of

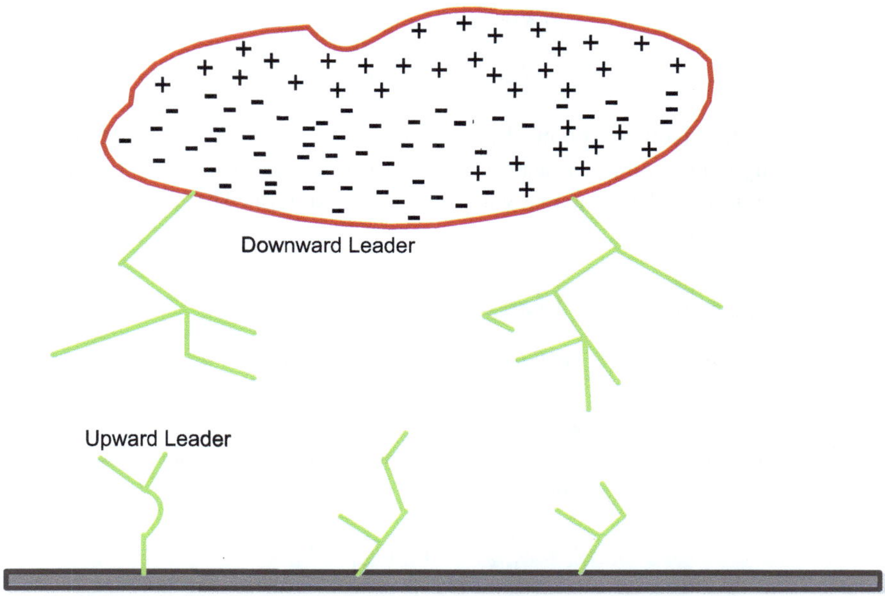


**Fig. 4.4** Formation of the charges and leaders in the cloud

lightning in nature. The lightning is based on four steps which are (1) breakdown of air at the surface of the cloud, (2) motion of the downward leader, (3) the motion of the upward leader, and (4) strike location. The mathematical representation of the LAPO algorithm can be depicted as follows:

**Step 1: Initialization**
The initial points or the trial spots of the downward leaders according to Eq. (4.25) are as follows:

$$X_{\mathrm{ts}}^{i} = X_{\min}^{i} + \left(X_{\max}^{i} - X_{\min}^{i}\right) \times \mathrm{rand} \tag{4.25}$$

where $X_{\mathrm{ts}}^{i}$ is the trial spots. $X_{\min}$ denotes the minimum boundary of the control variables, while $X_{\max}$ is the maximum value. Rand represents a random value within range [0,1]. Then, find the objective function of the initial points:

$$F_{\mathrm{ts}}^{i} = \mathrm{obj}\left(X_{\mathrm{ts}}^{i}\right) \tag{4.26}$$

**Step 2: The next jump determination**
The downward leader points are averaged and the objective function of this value is evaluated as follows:

$$X_{\mathrm{avr}} = \mathrm{mean}\ (X_{\mathrm{ts}}) \tag{4.27}$$

$$F_{\mathrm{avr}} = \mathrm{obj}\ (X_{\mathrm{avr}}) \tag{4.28}$$

$X_{\mathrm{avr}}$ denotes the averaged point. $F_{\mathrm{avr}}$ represents the averaged objective function point. For updating point $i$, a random solution $j$ is selected where $i \neq j$. Then, the updated solution will be compared with the selected solution. Then, the next jump is evaluated as follows.

$$X^i_{\mathrm{new}} = X^i_{\mathrm{ts}} + \mathrm{rand} \times \left(X_{\mathrm{avr}} + X^j_{\mathrm{PS}}\right) \qquad \mathrm{IF}\ \ F_j < F_{\mathrm{avr}} \tag{4.29}$$

$$X^i_{\mathrm{new}} = X^i_{\mathrm{ts}} - \mathrm{rand} \times \left(X_{\mathrm{avr}} + X^j_{\mathrm{PS}}\right) \qquad \mathrm{IF}\ \ F_j > F_{\mathrm{avr}} \tag{4.30}$$

**Step 3: Section Diminishing**

Section Diminishing mean pick the new solution if it better than the original solutionwhich can be considered as follows.

$$X^i_{\mathrm{ts}} = X^i_{\mathrm{new}} \qquad \mathrm{IF}\ \ F^i_{\mathrm{new}} < F^i_{\mathrm{ts}} \tag{4.31}$$

$$X^i_{\mathrm{new}} = X^i_{\mathrm{ts}}\ \mathrm{otherwise} \tag{4.32}$$

**Step 4: Motion of the upward leader**

In this step, the obtained Value from the previous solution is updated based on exponent operator, the best solution, and the worst solution as follows:

$$X^i_{\mathrm{new}} = X^i_{\mathrm{new}} + \mathrm{rand} \times S \times \left(A^i_{\mathrm{best}} - A^i_{\mathrm{worst}}\right) \tag{4.33}$$

where

$$S = 1 - \left(\frac{t}{t_{\max}}\right) \times e^{\left(-\frac{t}{t_{\max}}\right)} \tag{4.34}$$

where $t$ represents the current iteration, $t_{\max}$ represents the maximum number of iterations, and $X^i_{\mathrm{best}}$ and $X^i_{\mathrm{worst}}$ denote the best and the worst solution, respectively.

**Step 5: Strike point**

Strike point is the point of combination of the downward and upward leaders which represent the optimal solution. Flowchart of the LAPO with sensitivity analysis for allocation of PV and D-STATCOM is illustrated in Fig. 4.5.

The procedure of application the LAPO for optimal allocation the PV unit and D-STATCOM can be summarized as follows:

Step 1: Set the parameters of the LAPO as well as the system data.

Step 2: Run the load flow and carry out the sensitivity analysis of the system and assign the high potential buses for allocation the PV unit and D-STATCOM.

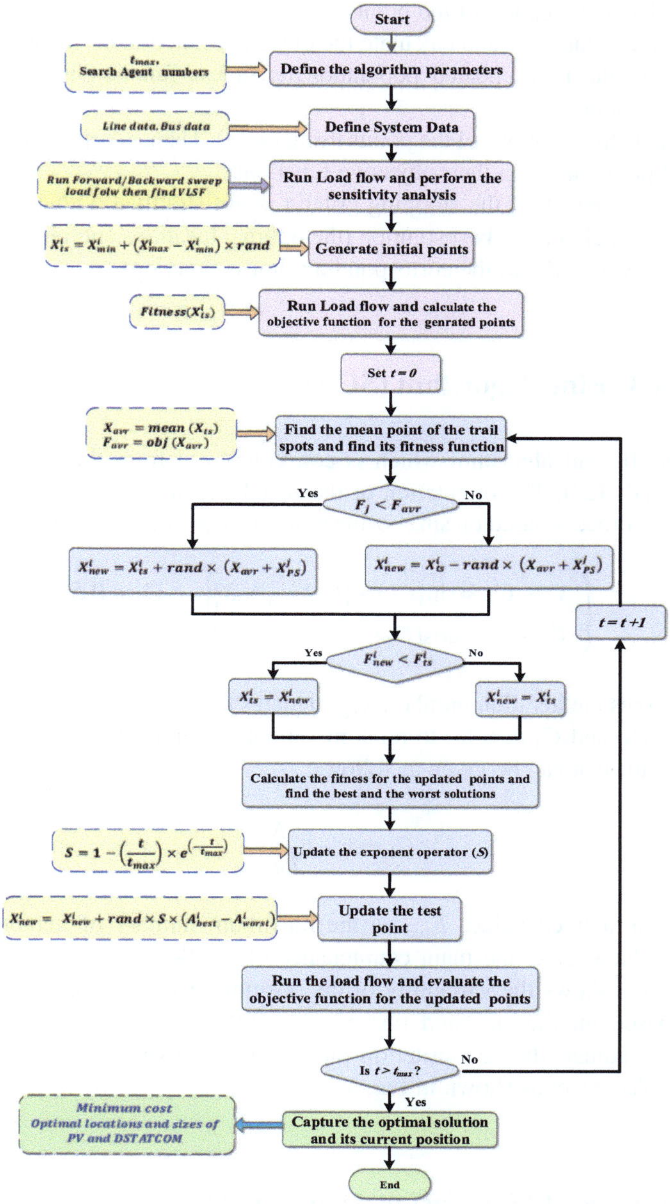


**Fig. 4.5** Solution process of LAPO and sensitivity analysis for allocation of PV and D-STATCOM

Step 3: Initialize the spot points randomly.

Step 4: Run the load flow and evaluate the objective function of the obtained points.

Step 5: Find the average vector point of the generated points and its corresponding objective function.

Step 6: Update the location of the points using (4.29) and (4.30), respectively.
Step 7: Run the load flow and calculate the objective function of the updated points and accept the new updated population its fitness function is better that the previous solution.
Step 8: Update the locations of the points using (4.33). Then, accept or reject the new updated point based on their objective functions.
Step 9: Go to step (4) if the stopping criteria is not stratified. Otherwise, end the program and obtain the best solution (the optimal locations and ratings of the PV and D-STATCOM and the corresponding fitness function).

## 4.4 Sine Cosine Algorithm (SCA)

SCA is an efficient algorithm which is conceptualized from the sine and cosine function trends [27]. The orientation of the search agents around the best solution with iterative process based on sine cosine trends is depicted in Eq. (4.35) as follows:

$$X_i^{t+1} = \begin{cases} X_i^t + C_1 \times \sin(C_2) \times \left|C_3 X_{\text{best}}^t - X_i^t\right| & C_4 < 0.5 \\ X_i^t + C_1 \times \cos(C_2) \times \left|C_3 X_{\text{best}}^t - X_i^t\right| & C_4 > 0.5 \end{cases} \tag{4.35}$$

where $t$ denotes the iteration number. $X_{\text{best}}^t$ denotes the best location from the search agents. $C_2$, $C_3$, and $C_4$ denotes to random variables within [0, 1]. $C_1$ represents an adaptive factor that can be given as follows:

$$C_1 = k - t\frac{K}{T_{\max}} \tag{4.36}$$

where $k$ is a fixed value. $T_{\max}$ is the maximum number of iterations. Equation (4.36) characterize the main countenance of the SCA algorithm as shown in Fig. 4.6 which shows the variability of the position of the groups around the best situation based on the size and phase angle difference of the sine and cosine functions. $y_1$ adjusts the new criterion placement to move outwardly or inwardly in the best placement as shown in Fig. 4.7.

### *4.4.1 Enhanced Sine Cosine Algorithm (ESCA)*

ESCA is based on Lévy flight distribution (LFD); LFD is integrated into SCA technology to foster the search ability and reconnaissance capacity of this optimization algorithm by rising the likelihood of manufacture new solutions to avoid algorithm stagnation and to avert trapping at local minimums. Lévy's trip is a random operation for generating a new solution based on a random walk whose

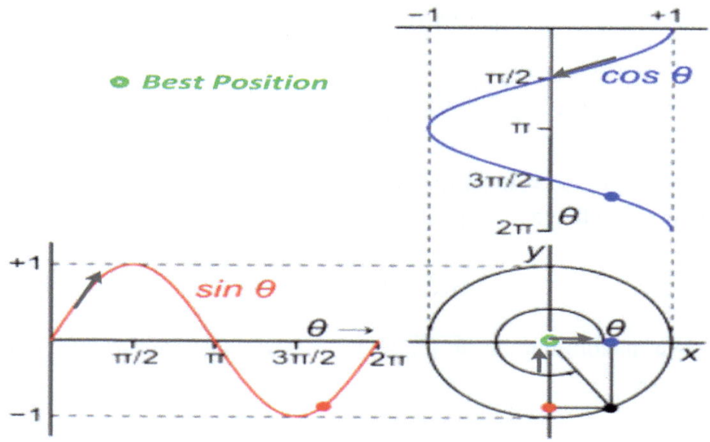


**Fig. 4.6** Inhabitance movement about the best solution based on cosine

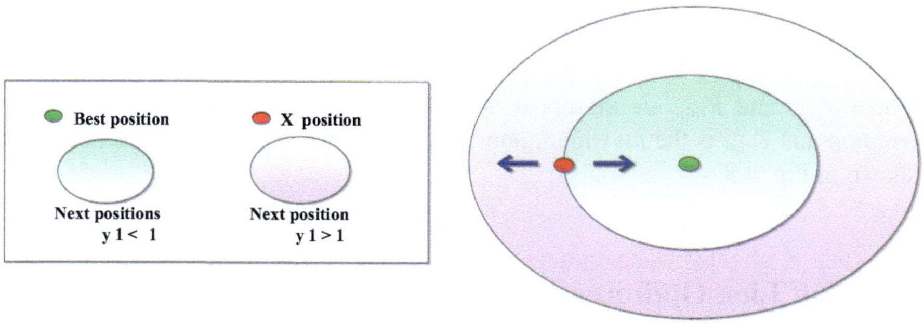


**Fig. 4.7** The trend of the following status on the best placement counts on $y_1$

steps are captured from the Lévy distribution. The new inhabitance situations which count on a fibrous distribution can be found as follows. ESCA is a new version of the SCA to overcome its stagnation. In the ESCA, the Lévy flight distribution (LDF) is utilized to improve the algorithm exploration ability where it enables the populations to jump to new areas. The new population position that is based on Lévy distribution can be found as follows [28]:

$$X_i^{\text{new}} = X_i + \propto \bigoplus \text{Levy}\ (\beta) \tag{4.37}$$

where $\propto$ denotes a random step size factor. $\bigoplus$ represents the multiplication of entry wise, while $\beta$ denotes an LDF. The step size is calculated given as:

$$\propto \bigoplus \text{Levy}(\beta) \sim 0.01 \frac{w}{|v|^{1/\beta}} \left(X_i^t - X_{\text{best}}^t\right) \tag{4.38}$$

where $u$ and $v$ are usually variables produced by a normal distribution where

$$w \sim n(0, \phi_u^2), v \sim n(0, \phi_v^2) \tag{4.39}$$

$$\phi_w = \left[\frac{\Gamma(1+\beta) \times \sin(\pi \times \beta/2)}{\Gamma[(1+\beta)/2] \times \beta}\right]^{1/\beta}, \phi_v = 1 \tag{4.40}$$

where $\Gamma$ is the criterion gamma function and $0 \leq \beta \leq 2$. To foster the utilization of SCA, the best search agent is updated by using changing band width as follows:

$$X_{\text{best}}^{\text{new}} = X_{\text{best}}^{t} \pm y_5 \times K_w \tag{4.41}$$

where $y_5$ is a random number in [0, 1]. $K_w$ is a variable bandwidth that decreases dynamically as:

$$K_w = K_{\max} e^{(E \times t)} \tag{4.42}$$

$$E = \left(\frac{\ln\left(\frac{K_{\min}}{K_{\max}}\right)}{T_{\max}}\right) \tag{4.43}$$

where $K_{\max}$ and $K_{\min}$ are the upper and lower sect width limits. t is the current iteration and $T_{\max}$ is the maximum number of iterations. The flowchart of ESCA is shown in Fig. 4.8.

## 4.5 Ant Lion Optimizer (ALO)

Ant lion optimizer (ALO) is a population-based optimization technique presented by Seydali Mirjalili in 2015 [146]. ALO simulates the hunting behavior of antlion and the interaction between the prey or the ants and the predator antlions where the antlions build circular traps to hunt the ants. The ants move in a stochastic pattern to search for their foods. The mathematic model of the stochastic movement of the ants is formulated as follows:

### *4.5.1 Random Movement of an Ant*

The random movement of the ant is described using the following equation:

$$X(t) = [0, \text{cumsum}(2r(t_1) - 1), \text{cumsum}(2r(t_2) - 1), \text{cumsum}(2r(t_n) - 1)] \tag{4.44}$$

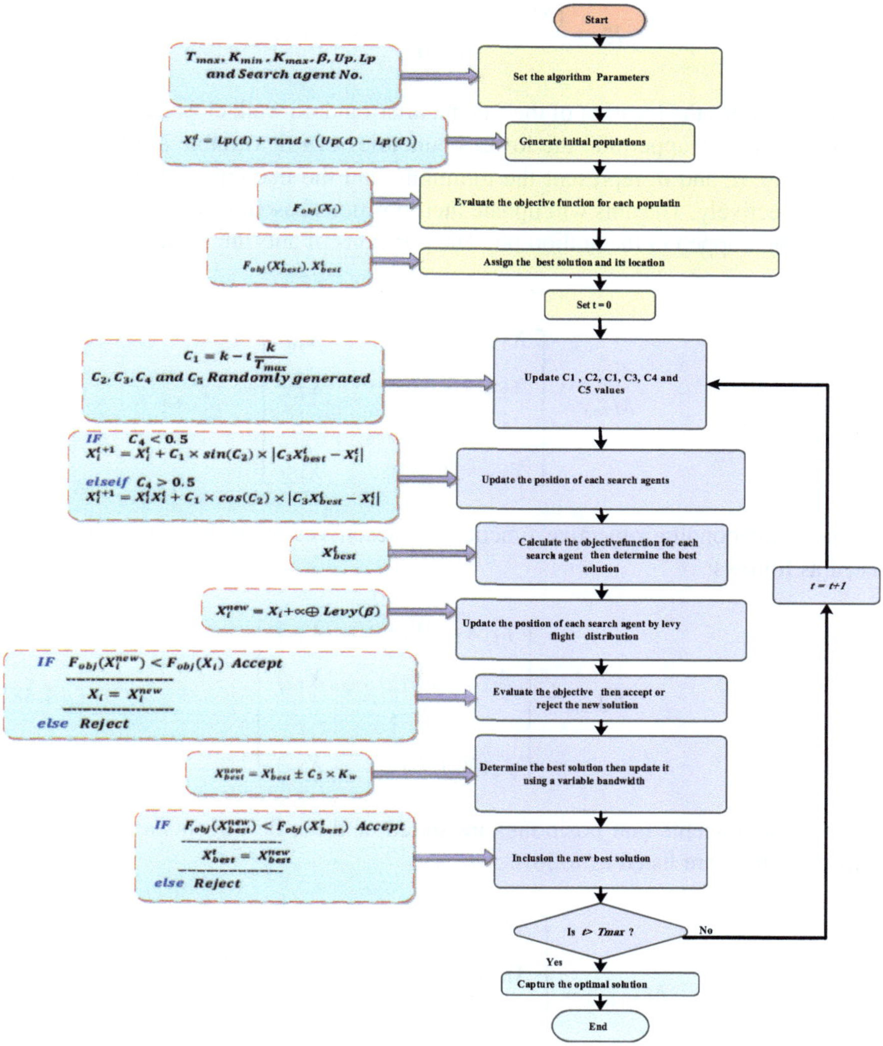


Fig. 4.8 Flowchart of ESCA

where $x(t)$ represents the location of the ant at the $t$-th iteration. cumsum denotes the cumulative sum. $N$ is the maximum number of iterations. $r(t)$ denotes a random which can be given as follows:

$$r(t) = \begin{cases} 1 & \text{if rand} > 0.5 \\ 0 & \text{if rand} \leq 0.5 \end{cases} \tag{4.45}$$

The min–max normalization equation is utilized for keeping the ants move in random walks inside the search spaces, which can be described as follows:

$$R_i^t = \frac{(X_i^t - a_i) \times (U_i^t - L_i^t)}{d_i - a_i} + L_i \tag{4.46}$$

where $R_i^t$ denotes the location of the $i$-th ant after random walk closed to $j$-th antlion. $U_i^t$ and $L_i^t$ are the upper and the lower boundaries of $i$-th variable at $t$-th iteration, respectively. $a_i$ and $d_i$ represent the minimum and the maximum steps of random walk, respectively. The ants will update their positions based on a random walk, and they will be trapped in the antlion pit. The positions of ants are listed in a matrix as follows:

$$M_{\text{Ant}} = \begin{bmatrix} X_{1,1} & X_{1,2} & \cdots & X_{1d} \\ X_{2,1} & X_{22} & \cdots & X_{2,1} \\ \vdots & \vdots & \ddots & \vdots \\ X_{n,1} & X_{n,2} & \ldots & X_{n,d} \end{bmatrix} \tag{4.47}$$

The corresponding objective functions for each vector of the ant positions are listed as follows:

$$\text{Obj}_{\text{Ant}} = \begin{bmatrix} f_1(X_{1,1}, X_{1,2}, \ldots, X_{1d}) \\ f_2(X_{1,1}, X_{1,2}, \ldots, X_{1d}) \\ \vdots \\ f_n(X_{1,1}, X_{1,2}, \ldots, X_{1d}) \end{bmatrix} \tag{4.48}$$

The search agents (ant positions) are sorted, and the best agents are selected as antlions, which are listed as follows:

$$M_{\text{Antlion}} = \begin{bmatrix} AL_{1,1} & AL_{1,2} & \cdots & AL_{1d} \\ AL_{2,1} & AL_{22} & \cdots & AL_{2,1} \\ \vdots & \vdots & \ddots & \vdots \\ AL_{n,1} & AL_{n,2} & \ldots & AL_{n,d} \end{bmatrix} \tag{4.49}$$

### 4.5.2 Trapping in Antlion Pits

The effects antlions' traps on the ant movement can be mathematically represented as follows:

$$L_i^t = \text{Antlion}_j^t + L_i^t \tag{4.50}$$

$$U_i^t = \text{Antlion}_j^t + U_i^t \tag{4.51}$$

### 4.5.3 Sliding Ants Toward Antlions

When the ants are trapped in the pit of the antlion, the upper and the lower bound should reduce with the increase of iteration as follows:

$$U_i^t = \frac{U_i^t}{I} \tag{4.52}$$

$$L_i^t = \frac{L_i^t}{I} \tag{4.53}$$

where $I$ represents a ratio that can be described as follows:

$$I = 10^{\omega} \frac{t}{T} \tag{4.54}$$

where $T$ and $t$ denote the maximum number of iterations and the current iteration. $\omega$ represents a constant which can be described as follows:

$$\omega = \begin{cases} 2 & t > 0.1T \\ 3 & t > 0.5T \\ 4 & t > 0.75T \\ 5 & t > 0.9T \\ 6 & t > 0.95T \end{cases} \tag{4.55}$$

### 4.5.4 Elitism

The elitism in optimization algorithm means the best solution (best antlion) is saved as an elite, which guide the motion of the populations in the iteration process and can be formulated as follows:

$$X_j^t = \frac{R_{\mathrm{A}}^t + R_{\mathrm{E}}^t}{2} \tag{4.56}$$

where $R_{\mathrm{A}}^t$ denotes the random walk closed to the best antlion using the roulette wheel for $t$-th iteration. $R_{\mathrm{E}}^t$denotes the position of randomly walking of the $j$-th ant, nearby the best or the elite antlion (E) in the swarm of ants.

### 4.5.5 *Catching Prey and Rebuilding the Pit*

The final stage of hunting behavior of antlions is catching an ant that reaches the pit's bottom. It must update its position to the latest position by the following equation:

$$\text{Antlion}_j^t = \text{Ant}_j^t \ \text{ if } f\left(\text{Ant}_j^t\right) > f\left(\text{Antlion}_j^t\right) \tag{4.57}$$

## 4.6 Modified Ant Lion Optimizer (MALO)

MALO is based on enhancing the basic ALO's searching capability by improving the exploration and the exploitation process. The exploration phase is enhanced by applying Lévy flight distribution (LFD), enabling the algorithm to jump to new areas to avoid the basic ALO's stagnation.

$$X_i^{\text{new}} = X_i + \propto \bigoplus \text{Levy}(\beta) \tag{4.58}$$

where $\propto$ denotes a random step parameter. $\bigoplus$ denotes the entry-wise multiplication. $\beta$ is a parameter related to the LFD? The step size is given as:

$$\propto \bigoplus \text{Levy}(\beta) \sim 0.01 \frac{u}{|v|^{1/\beta}} \left(X_i^t - \text{Antlion}_j^t\right) \tag{4.59}$$

where $u$ and $v$ denote variables obtained by normal distribution as.

$$u \sim N\left(0, \phi_u^2\right), \text{v} \sim N\left(0, \phi_v^2\right) \tag{4.60}$$

$$\phi_u = \left[\frac{\Gamma(1+\beta) \times \sin(\pi \times \beta/2)}{\Gamma[(1+\beta)/2] \times \beta}\right]^{1/\beta}, \phi_v = 1 \tag{4.61}$$

where $\Gamma$ represents the standard gamma function. $0 \leq \beta \leq 2$. The exploitation process of the algorithm is enhanced by updating the location of ants around the elite (best) solution in a spiral path as follows:

$$X_i^{\text{new}} = \left|\text{Antlion}_j^t - X_i\right| e^{bt} \cos(2\pi t) + \text{Antlion}_j^t \tag{4.62}$$

$b$ is a constant used to define the logarithmic spiral shape. To balance between the exploitation and the exploration, an adaptive operator is used for this sake as depicted in the following.

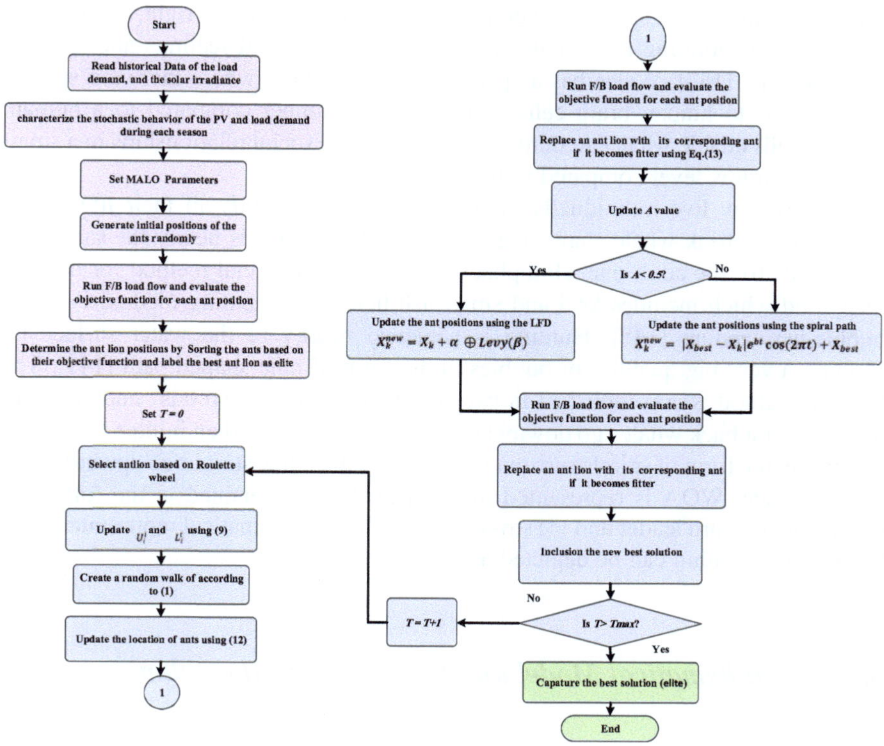


**Fig. 4.9** Flowchart of application of the MALO for optimal planning

$$A(t) = A_{\min} + \left(\frac{A_{\max} - A_{\min}}{T}\right) \times t \tag{4.63}$$

where $A_{\max}$ and $A_{\min}$ are the maximum and the minimum $A$ limits. This value is changed dramatically from $A_{\max}$ to $A_{\min}$ . When the value of $A$ is closed to$A_{\min}$, the position of the populations will be updated using (4.57) it while enhancing the exploration of this technique while when the value of $A$ is closed to $A_{\max}$, the position of the populations will be updated using (4.61), which enhance the exploitation of this technique. The application of the proposed algorithm for solving the planning problem is depicted in Fig. 4.9.

## 4.7 Whale Optimization Algorithm (WOA)

### *4.7.1 Inspiration*

The whale is considered the biggest animal on earth, even by its height or weight. Its length can exceed 30 meters and its weight can exceed 180 tons. Whales are categorized into seven species: killer, minke, blue, sei, humpback, right, and finback. It is worth mentioning that whales have high emotions and feelings above many other animals [147]. Wales has a special brain formation, where there are spindle cells similar to human brain cells with a double number compared to a human. Spindle cells enable whales to think, judge, and create social relationships in a smart way but at a low level compared to human [147].

Whales may live individually or in groups. Killer spends all their lifetime in groups. Humpback whales have huge baleen and their size is nearly the same as a school bus for the adult one. Humpback whales have a special method for hunting their food which includes krill and small fish herds; this hunting method is called bubble-net feeding [148]. Hunting is achieved nearby to the water surface by forming a very big quantity of bubbles in the form of nine-shaped path [149]. The hunting method is accomplished in two maneuvers "upward spirals" and "double-loops." Humpback whales go downward for nearly 12 meters then initiate producing bubbles in the form of spirals surrounding the prey then they start to go upward to the water surface. WOA is represented mathematically as explained in the following paragraph: upward leader and (4) strike location. The mathematical representation of the WOA algorithm can be depicted as follows:

## 4.7.2 *Mathematical Model and Optimization Algorithm*

Mathematical representation for hunting process can be explained as follows.

### 4.7.2.1 Circling Prey

The first step of the hunting process is to assign the position and surround the prey. Position of the prey is considered as the best solution which is nearby to the global solution within the search space. The mathematical equation representing this behavior can be formulated in the following equations.

$$\vec{D} = \left( \left| \vec{C} . \overrightarrow{X^*}(t) - \overrightarrow{X^*}(t) \right| \right) \tag{4.64}$$

$$\vec{X}(t+1) = \overrightarrow{X^*}(t) - \vec{A} . \vec{D} \tag{4.65}$$

$t$ denotes the current iteration

$A$ and $C$ represent vectors coefficient

$X^*$ represents the position of best solution

$X$ represents the position vector

The resulted of the best solution $X^*$ is characterized by updating its value for each iteration until reaching to global optima or the best solution.

$\vec{A}$ and $\vec{C}$ can be calculated from the following equations.

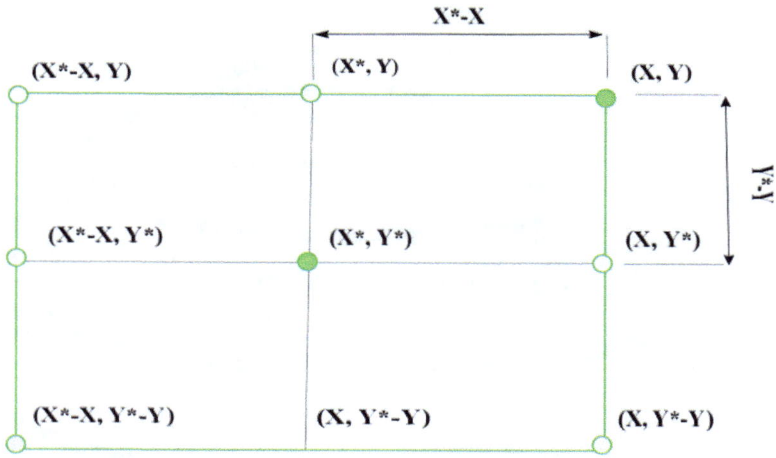


**Fig. 4.10** Position vectors in 2D along with the expected positions

$$\vec{A} = 2\vec{a}.\vec{r} - \vec{a} \tag{4.66}$$

$$\vec{C} = 2.\vec{r} - \vec{a} \tag{4.67}$$

where $\vec{a}$ is a coefficient that decreases its value within the search space from 2 to 0 through proceeding of the process, and $\vec{r}$ is a vector that varies randomly within the period [0,1]. Fig. 4.10 shows the 2D relationship between the current positions of candidate $(X, Y)$ and the best record $(X^*, Y^*)$ according to vectors $\vec{A}$ and $\vec{C}$ values.

Equation (4.65) represents the mathematical formulation of encircling the prey by the search agents inside the search space. For $n$ dimensions search space, the same idea can be applied. The candidates are moving in hypercube around the best solution. The second stage of the hunting process is explained as follows:

#### 4.7.2.2 Bubble-Net Attacking Method

This method of humpback whales for hunting process represents the exploitation phase can be applied according to two mechanisms; this foraging is done by creating distinctive bubbles along a circle or "nine"-shaped path as shown in Fig. 4.11, which forms a curtain and frights the prey for getting them together to speed up the efficiency of hunting [148, 150].

#### 4.7.2.3 Shrinking Encircling Mechanism

In this mechanism, the vector $\vec{a}$ decreases according to (4.66). The vector $\vec{A}$ changes its value within the interval $[-a, a]$ which descending from 2 to 0. If the

**Fig. 4.11** Spiral bubble-net behavior of humpback whales

interval of $\vec{A}$ is set to [−1, 1], the search agents will update their locations between the original position and the new location of the best agent, the relative locations of the (*X*, *Y*) with respect to (*X*∗, *Y*∗) lies within $0 \le A \le 1$ when the search space is 2D.

#### 4.7.2.4 Spiral Updating Position

The spiral path of humpback whales is depicted in the mathematical equations of this mechanism and can be formulated as follows.

$$\vec{X}(t+1) = \vec{D'}.e^{bl}.\cos(2\pi l) + \overrightarrow{X^*}(t) \tag{4.68}$$

$$\overrightarrow{D'} = \left|\overrightarrow{X^*}(t) - \vec{X}(t)\right| \tag{4.69}$$

where $D'$ is the distance from the prey to the humpback whale (best solution), *b* is the shape coefficient for the spiral movement, *l* is a number vary randomly within the interval [−1, 1], and "." represents that to multiply element by element. The spiral movement of hunting process is reduced in the form of two shapes in the same time. It can be considered as a probability of 50% for choice between the spiral shape and the encircling mechanism as follows.

$$\vec{X}(t+1)=\begin{cases}\overrightarrow{X^*}(t)-\vec{A}.\vec{D} & \text{if } p<0.5\\ \vec{D'}.e^{bl}.\cos(2\pi l)+\overrightarrow{X^*}(t) & \text{if } p\geq 0.5\end{cases} \tag{4.70}$$

The subscript $p$ is a number vary randomly within the interval [−1, 1]. The humpback whales' hunting process can be achieved randomly beside the bubble net method. Random search is explained in the following.

#### 4.7.2.5 Search for Prey

Search for the prey represents the exploration phase based on changing the value of vector $\vec{A}$. Random search of the humpback whales depends upon the position of the whales with respect to each other. The value of vector $\vec{A}$ can be changed around 1.0 to oblige the prey to start moving far from the reference whale. The exploitation phase depends upon updating the position of search agents according to the best agent obtained so far. However, the exploration phase depends upon updating the position of search agents randomly. According to this process and $\vec{A}$ >1, exploration phase is ensured and the WOA algorithm can reach to global optima. Representation of this model is formulated in the following eqs. [150].

$$\vec{X}(t+1)=\left|\vec{C}.\overrightarrow{X_{\text{rand}}}(t)-\vec{X}\right| \tag{4.71}$$

$$\vec{X}(t+1)=\overrightarrow{X_{\text{rand}}}-\vec{A}.\vec{D} \tag{4.72}$$

The subscript $\overrightarrow{X_{\text{rand}}}$ denotes a random position vector that refers to randomly chosen whale from the search agents (Fig. 4.12).

## 4.8 Slime Mold Algorithm (SMA)

This algorithm simulates a method for finding multiple heads in a velarium that inhabits cold and wet places. In this technique, weights approach to negative and positive feedback generated by sticky mold during the foraging process. In this stage, the slurry can determine the best food-gathering path in a superior way. The organic matter in the slime mold searches for food. Then, it encircles it and releases enzymes for its consumption. In the migration stage, the anterior end expands into a fan shape along with a venous network that allows the cytoplasm to slide inward. An intravenous network consists of using multiple food sources simultaneously to form a connection between them. In this mechanism, a reproductive wave is formed when a vein approaches a food source. The mathematical representation of MSA is formulated as follows:

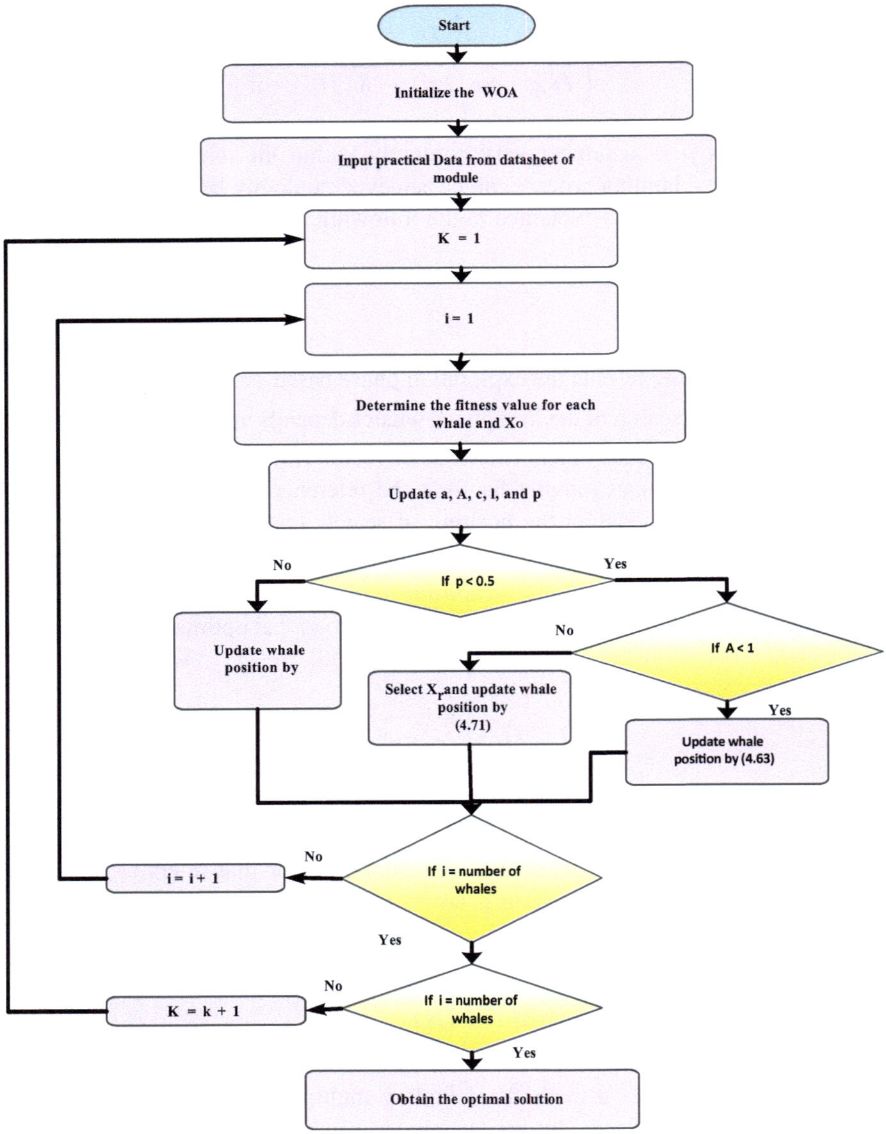


**Fig. 4.12** Flowchart of WOA optimization technique

## *4.8.1 Approach Food*

The shrinkage mode of a slime mold can be represented as follows [151]:

$$X(t+1) = \begin{cases} X_b(t) + vb \cdot (W \cdot X_A(t) - X_B(t)), r < p \\ vc \cdot X(t), r \geq p \end{cases} \tag{4.73}$$

where $vb$ denotes a parameter with a range of $[-a, a]$, $vc$ is reduced from one to zero, $t$ denotes the current iteration, $X_b$ is the location of slime mold with the highest odor concentration currently assigned, $X$ denotes the location of slime mold, $X_A$ and $X_B$ are two individuals randomly selected from the swarm, and $W$ represents the weight of slime mold. The $p$ is a parameter which can be obtained as follows:

$$p = \tan h|S(i) - \mathrm{DF}| \tag{4.74}$$

where $i \in 1, 2, \ldots, n$, $S(i)$ represents the fit of $\vec{X}$ and DF represents the best fit obtained in all iterations. The formula of $\vec{vb}$ is as follows:

$$vb = [-a, a] \tag{4.75}$$

$$a = \arctan h\left(-\left(\frac{t}{\max\_t}\right) + 1\right) \tag{4.76}$$

The $W$ is depicted using (4.77) as follows:

$$W = \begin{cases} 1 + r \cdot \log\left(\dfrac{bF - S(i)}{bF - wF} + 1\right), \text{condition} \\ 1 - r \cdot \log\left(\dfrac{bF - S(i)}{bF - wF} + 1\right), \quad \text{others} \end{cases} \tag{4.77}$$

$$\mathrm{SmellIndex} = \mathrm{sort}(S) \tag{4.78}$$

where condition indicates that $S(i)$ classifies the first half of the population, $r$ denotes the random value in the interval of [0, 1], bF denotes the optimal fitness obtained in the current iterative process, and $wF$ denotes the worst fit it value that has been getting it in the iterative process. Currently, smell index refers to a sequence of ranked fitness values (escalating into the minimum value problem).

### 4.8.2 Wrap Food

The case simulates a slime mold to control research patterns related to food quality. If the food concentration is contained, the weight near the area is greater; when the concentration of food is low, the weight of the area will decrease, and thus, it will turn to explore other areas. Figure 4.13 illustrates the process of evaluating the fit values of a slime mold. Based on the principle above, the mathematical formula for updating a slime mold site is as follows.

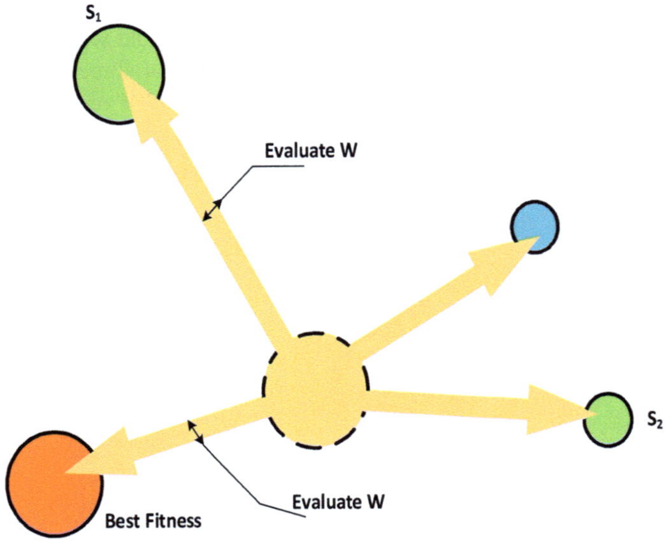


**Fig. 4.13** Evaluation of fitness

$$X^* = \begin{cases} \text{rand} \cdot (\text{UB} - \text{LB}) + \text{LB}, \text{rand} < z \\ X_b(t) + vb \cdot (W \cdot X_\text{A}(t) - X_\text{B}(t)), r < p \\ vc \cdot X(t), r \geq p \end{cases} \tag{4.79}$$

where LB and UB are the bounds of the lower and upper control variables, and rand and $r$ denote the random value in [0, 1].

### *4.8.3 Grabble Food*

The value of $vb$ randomly varied within range $[-a, a]$ and gradually reached zero as the iteration number increases. The value of $vc$ varied between [−1, 1] and reached zero eventually (Fig. 4.14).

**Algorithm 1 Pseudo-code of SMA**

Define the parameters of SMA the maximum number of iterations.

Generate initial positions of slime mold $X_i(i = 1, 2, \ldots, n)$;

While ($t \leq$ *Max _ iteraition*)

Evaluate the objective function of all slime mold.

*Update bestFitness*, $X_b$

Obtain the W using Eq. (4.77).

For *each search portion*

*Update p, vb, vc*;
*Update locations by* Eq. (4.79);

End For
$t = t + 1$;
End while
*Return bestFitness*, $X_b$;

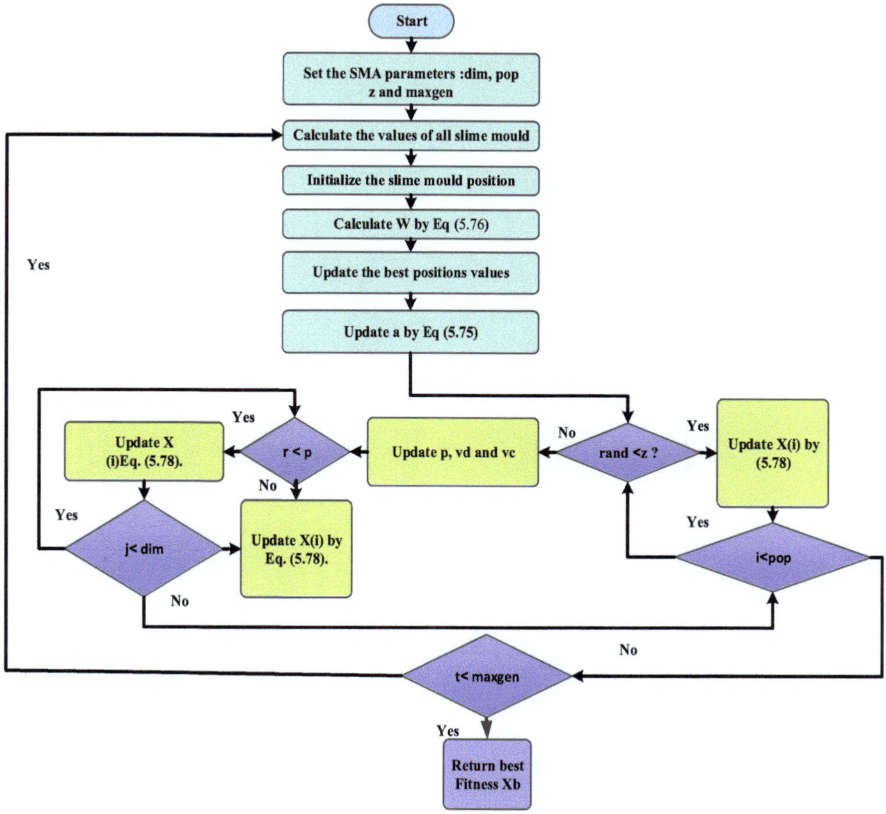


**Fig. 4.14** Flowchart of SMA optimization technique

# Chapter 5
# Results and Discussion

## 5.1 IEEE-30 Bus Radial Distribution System

In this chapter, the numerical results are presented where the locations and sizes of PV units and D-STATCOMs are determined for the East Delta Network (EDN) as a part of the Unified Egyptian Network (UEN) grid. The system data for the EDN grid is found in [23]. The schematic diagram of the (EDN) grid is shown in Fig. 5.1. The system load demand is 11 kV. The system active and reactive load demands are 22441.259 kW and 14162.265 kVar, respectively. The minimum voltage magnitude equals 0.94626 p.u. at bus 30. The total active loss is 805.733 kW. Lines and bus data for this system are given in Appendix B.

### *5.1.1 Optimal Allocation of PV and D-STATCOM in RDS Using ALO Algorithm*

The selected parameters of ant lion optimizer (ALO) and the system constraints are listed in Table 5.1. The studied cases are listed in Table 5.2, which include the optimal integration of PV units, installation of D-STATCOMs only, and installation of PV units along with D-STATCOMs. Table 5.2 shows that the active power losses are reduced from 805.733 (base case) to 601.782 kW, 319.228 kW, and 113.620 kW with optimal integration of D-STATCOMs only, installation of PV units only, and installation of PV units along with D-STATCOMs, respectively. The voltage profile is enhanced from 1.0669 p.u. (base case) to 0.8288 (with D-STATCOM only) p.u. 0.1301 p.u. (with PV only) and 0.0651 p.u. (PV with D-STATCOM). Figure 5.2 shows the voltage profile of the system, which shows that the inclusion of PV along with D-STATCOM can enhance the voltage profile considerably. Table 5.2 shows



A. M. A. El-Hameid et al., *Enhancement of Grid-Connected Photovoltaic Systems Using Artificial Intelligence*, https://doi.org/10.1007/978-3-031-29692-5_5

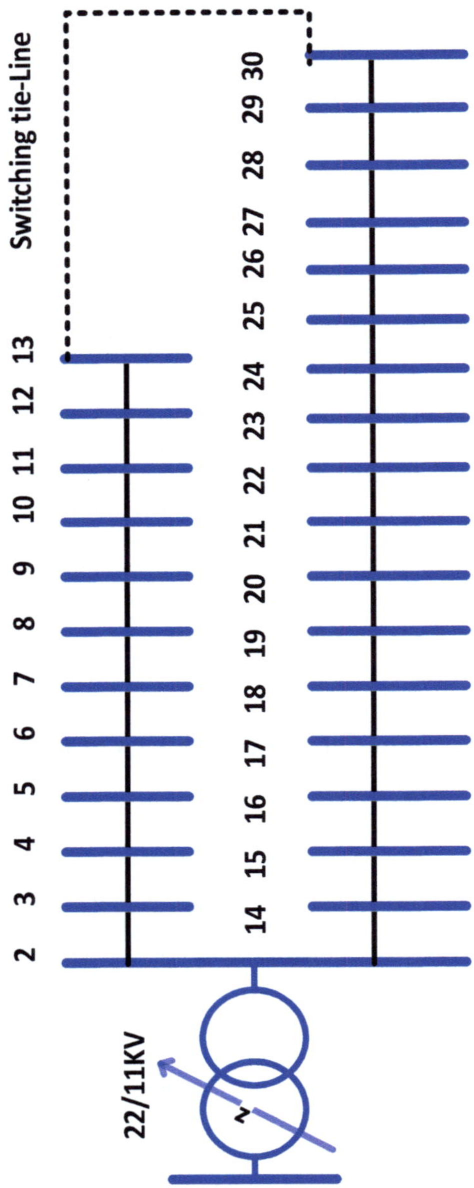


**Fig. 5.1** Single line diagram of the (EDN)

**Table 5.1** Parameters of system

| Parameter | Value |
|---|---|
| Maximum iteration | 100 |
| Population number | 30 |
| Voltage limits | $0.90 \leq V_i \leq 1.05$ p. u. |
| PV sizing limits | $0 \leq P_{DG} \leq 23$ MW |
| D-STATCOM sizing limits | $0 \leq Q_{STATCOM} \leq 14$ MVar |

**Table 5.2** Simulation results with or without PV units and D-STATCOM

| Values | Basic | With D-STATCOM | With PV | PV with D-STATCOM |
|---|---|---|---|---|
| Optimal PV location (size kW) | – | – | 19 (14070)<br>7 (7434) | 6 (80221)<br>20 (11546) |
| Optimal D-STATCOM location (size kVar) | – | 5 (5626.6)<br>19 (7973) | – | 20 (57948)<br>14 (6063) |
| Total $P_{PV}$ (kW) | – | – | 21,504 | 917,677 |
| Total $Q_{DSTATCOM}$ (KVar) | – | 13599.6 | – | 64,011 |
| $V_{min}$ (p. u.) @ bus No. | 0.94626 @ bus 30 | 0.95707 @ bus 30 | 0.99006 @ bus 30 | 0.99570 @ bus 30 |
| Total $P_{losses}$ (kW) | 805.733 | 601.782 | 319.228 | 113.620 |
| Total $Q_{loss}$ (kVar) | 61.184 | 267.062 | 134.715 | 42.563 |
| VD (p. u.) | 1.0669 | 0.8288 | 0.1301 | 0.0651 |
| ∑VSI | 24.9867 | 25.8400 | 28.48 | 28.8539 |

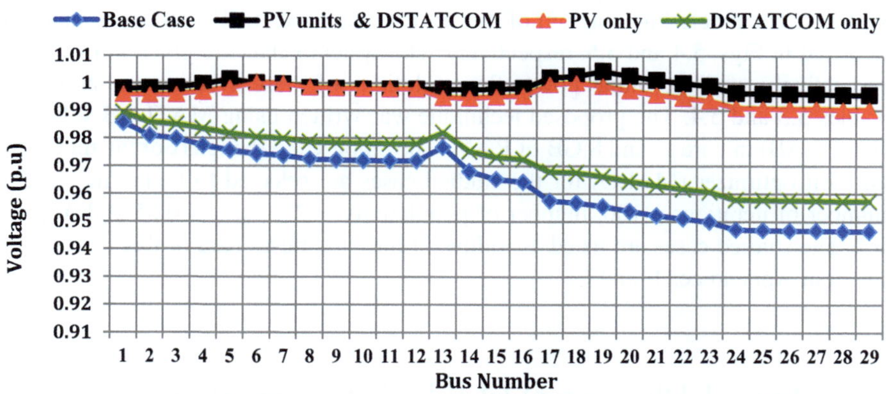


**Fig. 5.2** Voltage profile of the system

that the voltage stability also enhanced the inclusion of PV along with D-STATCOM. Figure 5.13 shows the voltage stability index (VSI) of the system; the VSI was enhanced considerably with the inclusion of PV units along with D-STATCOM compared with other tasks.

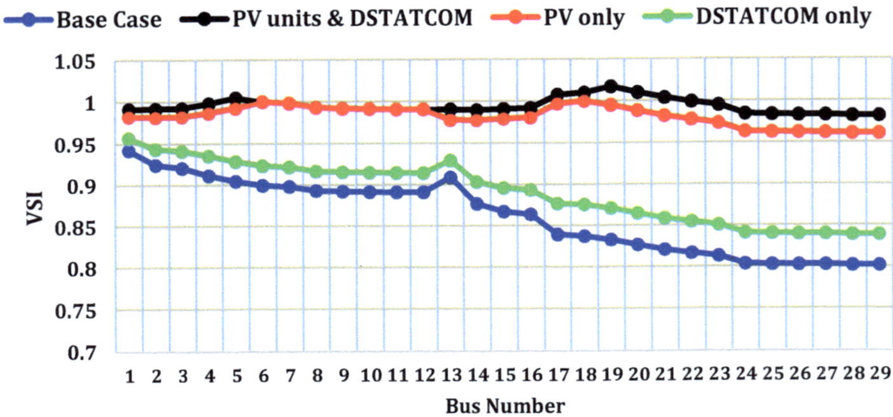


**Fig. 5.3** Voltage Stability index of the system

## 5.2 IEEE-69 Bus and IEEE-118 Bus Radial Distribution System

In this section, the proposed algorithm is applied for solving the optimal power problem for optimal allocation of a hybrid system (PV-DG and the D-STATCOM) under uncertain conditions. Two test systems are utilized to the optimal planning of incorporating the hybrid system which includes IEEE-69 bus and IEEE-118 bus systems. The single line diagram of the IEEE-69 bus and IEEE-118 bus systems is depicted in Figs. 5.4 and 5.5, respectively. The system's data are given in [57, 152] respectively, and the initial power flow solutions of these systems are shown in Tables 5.3 and 5.4. The proposed technique was written using MATLAB software (MATLAB) in core I7 and 8 GB of RAM. The empirical parameters of the modified ant lion optimizer (MALO) for the studied cases are selected to be the maximum number of iterations =100, number of populations = 25, $A_{max}$ = 0.85 and $A_{min}$ = 0.4. The studied cases are listed as follows: Lines and buses data of this system are given in Appendices E and C.

## 5.3 Optimal Allocation of DER in RDS Using ALO and MALO

### *5.3.1 Case 1: Optimal Installation of PV System Under the Deterministic Condition*

In this case, to state the ALO and the MALO's effectiveness, these techniques have been examined on the standard IEEE-69 bus to assign the optimal locations and

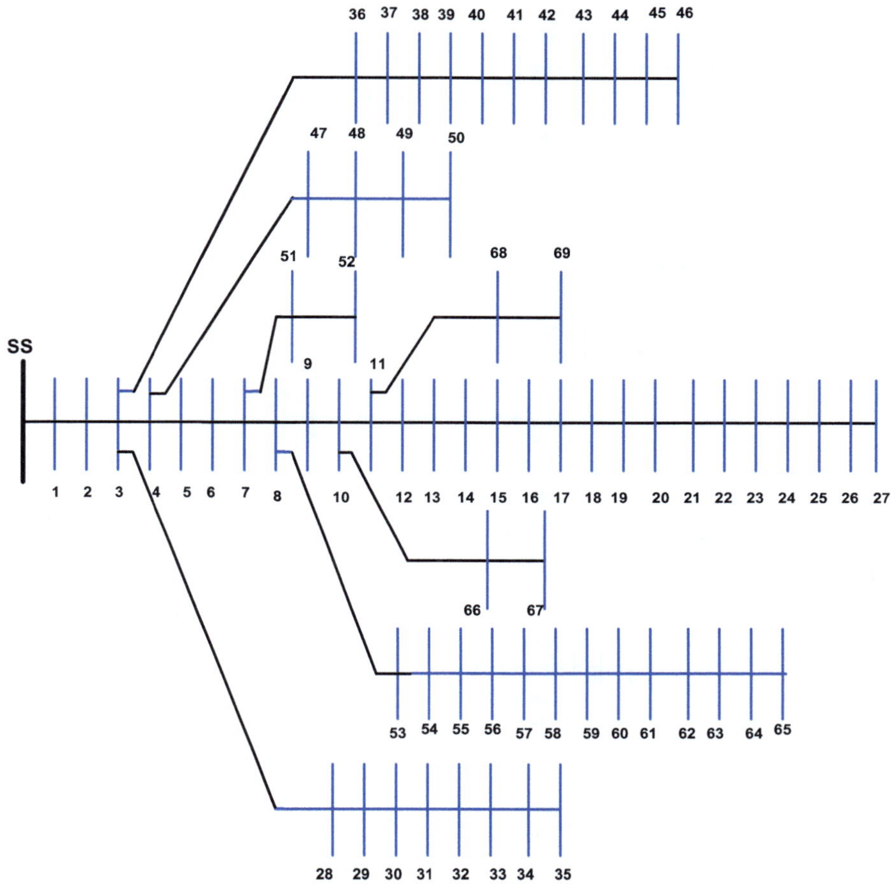


**Fig. 5.4** Single line diagram of the IEEE-69 bus

ratings of the PV units for power loss minimization. Single, two, and three PV units are installed optimally, and the obtained results are compared with other reported techniques. The obtained results are listed in Table 5.5. Judging from this table, the power losses are reduced considerably with the increasing number of the installed PV units. Referring to the comparison of Table 5.6, it clear that the obtained power losses by MALO are better than ALO and the reported algorithms, which verifies the effectiveness of the proposed algorithm. The convergence characteristics of the MALO are shown in Fig. 5.6. This figure illustrates that the proposed algorithm has excellent and stable convergence characteristics where there is no oscillation appeared.

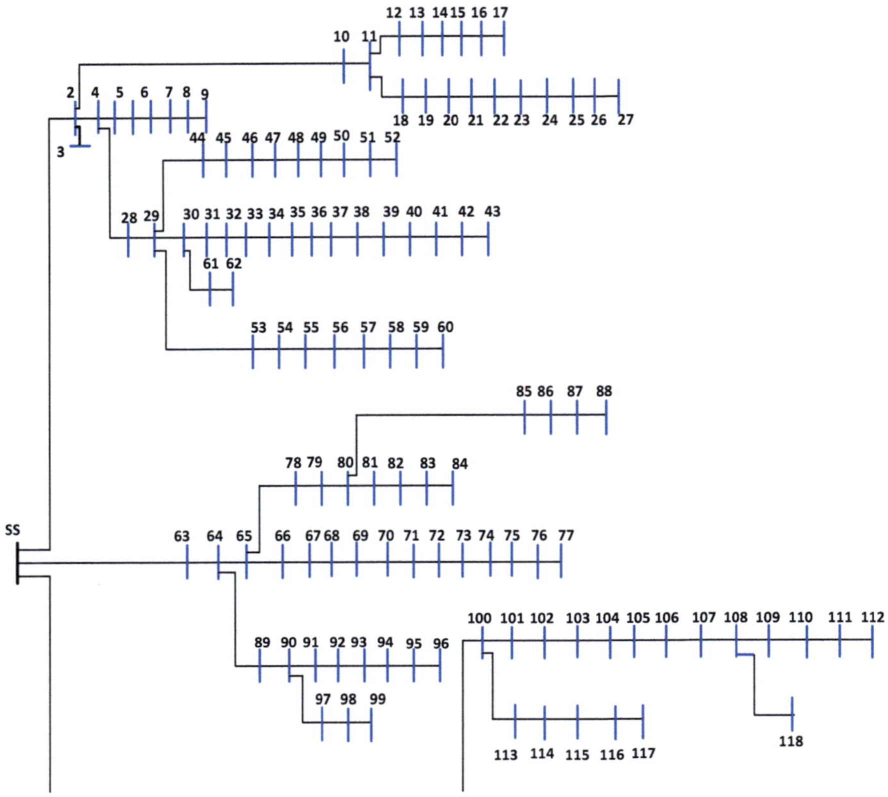


**Fig. 5.5** Single line diagram of 118-bus system

**Table 5.3** The system specification of 69-bus systems and its initial power flow

| Item | Value |
|---|---|
| *System specifications:* | |
| NB | 69 |
| Npr | 68 |
| Vsys (kV) | 12.66 |
| Base MVA | 100 |
| $S_{\text{Load}}$(MVA) | 3.802 + j2.694 |
| $P_{\text{Total}_{\text{loss}}}$ (kW) | 225 |
| $Q_{\text{Total}_{\text{loss}}}$ (kVar) | 102.198 |
| $V_{\min}$ (p. u.) @ bus | 0.90919, 65 |

**Table 5.4** The system specification and initial power flow

| Item | Value |
|---|---|
| *System specifications:* | |
| NB | 118 |
| Npr | 117 |
| Vsys (kV) | 12.66 |
| Base MVA | 100 |
| $S_{\text{Load}}$(MVA) | 22.7097 + j17.0412 |
| $P_{\text{Total}_{\text{loss}}}$ (kW) | 1298.091 |
| $Q_{\text{Total}_{\text{loss}}}$ (kVar) | 978.797 |
| $V_{\min}$ (p. u.) @ bus | 0.86880@ 77 |

**Table 5.5** Results of optimal allocation of PV units for loss reduction

| Items | Without PV | With PV | | | | | |
|---|---|---|---|---|---|---|---|
| | | One PV | | Two PVs | | Three PVs | |
| | | MALO | ALO | MALO | ALO | MALO | ALO |
| Total losses (kW) | 225 | 83.222 | 83.222 | 71.674 | 72.518 | 69.426 | 71.714 |
| Minimum voltage | 0.9091 @ bus 65 | 0.9682 @ bus 27 | 0.9683 @ bus 27 | 0.9789 @ bus 65 | 0.9789 @ bus 65 | 0.9789 @ bus 65 | 0.978 @ bus 65 |
| Maximum voltage | 0.999 @ bus 2 | 0.999 @ bus 2 | 0.9999 @ bus 2 | 0.999 @ bus 2 | 0.9999 @ bus 2 | 0.999 @ bus 2 | 1.000 @ bus 40 |
| PV size (location kW) | – | 1872 (61) | 1872 (61) | 531 (17) 1781 (61) | 813.94 (12) 1735 (61) | 1719 (61) 523 (11) 381 (18) | 528.8 (18) 380 (39) 1780 (61) |
| VSI (p.u.) | 61.2181 | 64.6212 | 64.6212 | 66.0295 | 65.7845 | 66.224 | 66.0396 |
| VD (p.u.) | 1.8374 | 0.8729 | 0.8729 | 0.5002 | 0.5647 | 0.4504 | 0.4980 |

### 5.3.2 *Case 2: Optimal Planning Under Uncertainties of System*

#### 5.3.2.1 IEEE-69 Bus System

In this case, the optimal planning problem is solved using the proposed algorithms under the uncertainties of the connected load and solar irradiances on IEEE-69 bus system. The optimal planning problem is solved for a multi-objective function, including the cost reduction, the voltage profile, and stability index improvement. It should be pointed out here that 3 years of hourly historical data of load demand

**Table 5.6** Comparative results for incorporating PV in the 69-bus system

| Type | Technique | Optimal sizes (kW) | Locations | Power loss (kW) |
|---|---|---|---|---|
| | Without | – | – | 225 |
| 1 PV | Analytical [153] | 1807.8 | 61 | 92 |
| | Exhaustive OPF [154] | 1870 | 61 | 83.23 |
| | EA-OPF [154] | 1870 | 61 | 83.23 |
| | MTLBO [155] | 1819 | 61 | 83.323 |
| | CSA [156] | 2000 | 61 | 83.8 |
| | Hybrid [157] | 1810 | 61 | 83.372 |
| | PSO [157] | 1870 | 61 | 83.8 |
| | GA [158] | 1794 | 61 | 83.4252 |
| | ALO | 1872.7 | 61 | 83.222 |
| | MALO | 1872.7 | 61 | 83.222 |
| 2PV | MINLP [159] | 510<br>1780 | 17<br>61 | 71.693 |
| | GA [160] | 1777<br>555 | 61<br>11 | 71.7912 |
| | CSA [156] | 600<br>2100 | 22<br>61 | 76.4 |
| | SGA [156] | 1000<br>2400 | 17<br>61 | 82.9 |
| | PSO [156] | 700<br>2100 | 14<br>62 | 78.8 |
| | MTLBO [155] | 519.705<br>1732.004 | 17<br>61 | 71.776 |
| | GA [158] | 886<br>861 | 61<br>62 | 84.233 |
| | ALO | 813.947<br>1735 | 12<br>61 | 72.518 |
| | MALO | 531.4925<br>1781 | 17<br>61 | 71.674 |
| 3PV | EA [154] | 467<br>380<br>1795 | 11<br>18<br>61 | 69.62 |
| | MTLBO [155] | 493<br>378<br>1672 | 11<br>18<br>61 | 69.539 |
| | KHA [161] | 496<br>311<br>1735 | 12<br>22<br>61 | 69.56 |
| | Hybrid [157] | 510<br>380<br>1670 | 11<br>17<br>62 | 69.52 |
| | PSO [157] | 460<br>440<br>1700 | 11<br>17<br>61 | 69.541 |

(continued)

**Table 5.6** (continued)

| Type | Technique | Optimal sizes (kW) | Locations | Power loss (kW) |
|---|---|---|---|---|
| | EA [154] | 467<br>380<br>1795 | 11<br>18<br>61 | 69.62 |
| | ALO | 528.8338<br>380<br>1780.9 | 18<br>39<br>61 | 71.714 |
| | MALO | 1719.0<br>523<br>381.3461 | 61<br>11<br>18 | 69.426 |

and solar irradiance have been considered in this chapter. Based on this, every single year has been split into four seasons. To characterize the stochastic behavior of the PV and load demand during each season, a day within that season (24 − h) is considered. Thus, each year has 96 time periods (4 seasons × 24 − h). The obtained load profiles and solar irradiance under uncertain conditions are depicted in Figs. 5.7 and 5.8.

At the base case (without the inclusion of PV or D-STATCOM), the total cost, TVD, and TVSI are 1.24867E+4 p.u., 5.48787E+5 p.u., and 2.70673E+6 $, respectively. The optimal planning problem is solved with the inclusion of single and two PV-DGs and DTSTACOMs using the proposed algorithm. The cost data of the PV-DG and the D-STATCOM are listed in Table 5.7. The simulation results for optimal integration of a single and two hybrid PV-DG and the D-STATCOM are listed in Tables 5.8 and 5.9, respectively.

In the inclusion of a single hybrid system, the total cost is reduced considerably to 2.5012E+06 $ or by 7.59% compared to without insertion PV-DG or D-STATCOM as well as the TVD is reduced to 6.0753E+3 p.u. (51.35%) and the TVSI is enhanced to 5.7645E+5 p.u. (5.04%). Besides that, the drawn energy from the grid and the energy losses are reduced considerably. The hybrid system's assigned best location is at the 58th bus, while the size of the PV-DG and D-STATCOM is 3389 kW and 2691 kVAR, respectively.

The output power of the PV unit is depicted in Fig. 5.9. It is evident that the output power of the PV unit follows the changes in the solar irradiance. Judging from Table 5.8, the obtained results by the proposed algorithm are better than those obtained the conventional ALO, SCA, WOA, and GOA in terms of the cost, **TVD**, and **TVSI**.

In this case, two-hybrid systems are embedded in the IEEE-69 bus distribution system. According to Table 5.9 the total cost is reduced considerably to 2.41485E+6 $ or by 10.78% compared to base case, and the **TVD** is reduced to 5.300667E+3 p.u. (57.55%), and the **TVSI** is increased to 5.801994E+5 p.u. (5.72%). The assigned best locations of the hybrid systems for this case are at the 61st bus and the 12th bus, while the sizes of the first PV-DG and D-STATCOM are 1646 kW and

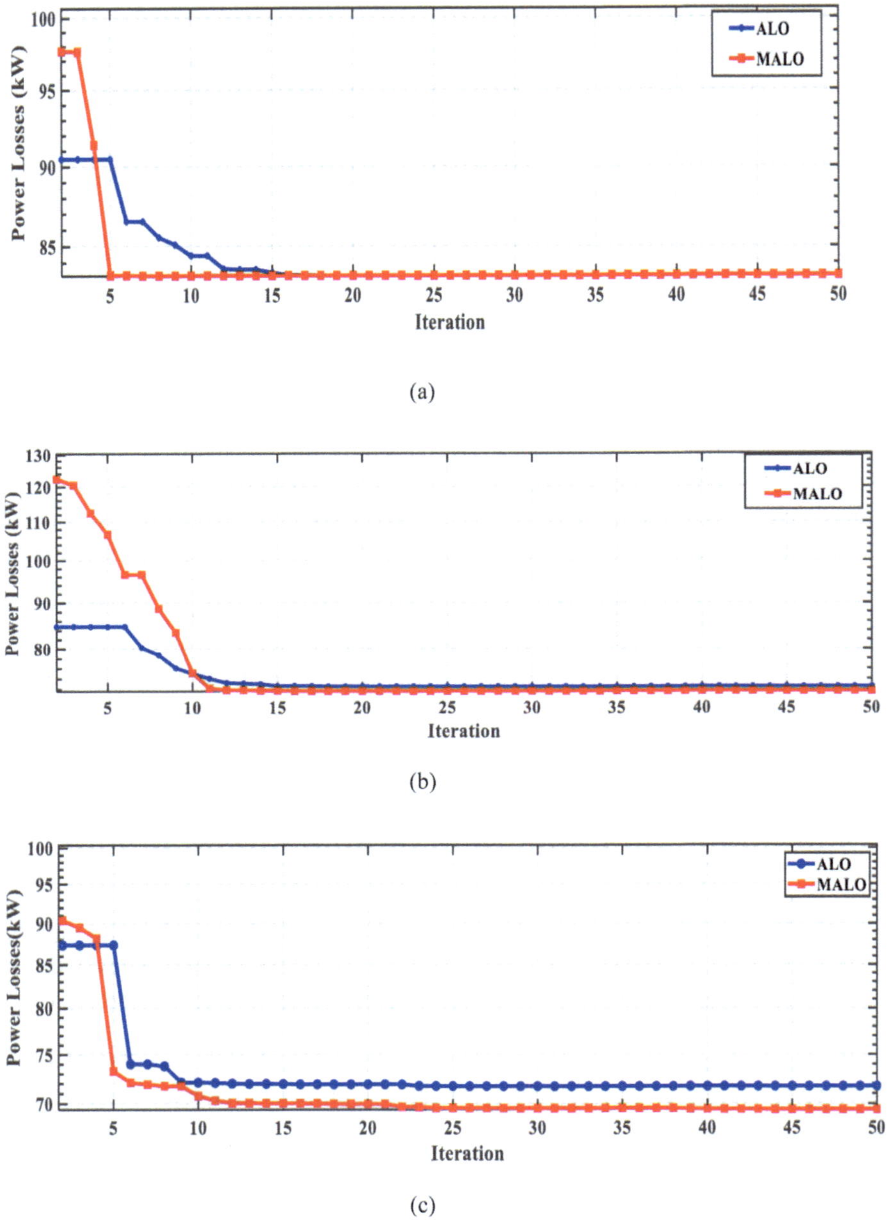


**Fig. 5.6** The convergence characteristic of the ALO and MALO for power losses minimization with incorporating (**a**) single PV unit, (**b**) two PV units, and (**c**) three PV units

1987 kVAR, respectively. The sizes of the second PV-DG and D-STATCOM are 2155 kW and 707 kVAR, respectively. The first and second PV units' output power are depicted in Figs. 5.10 and 5.11, respectively. Referring to Figs. 5.10 and 5.11, the output power of PV units is varied during day ahead with solar irradiance variations.

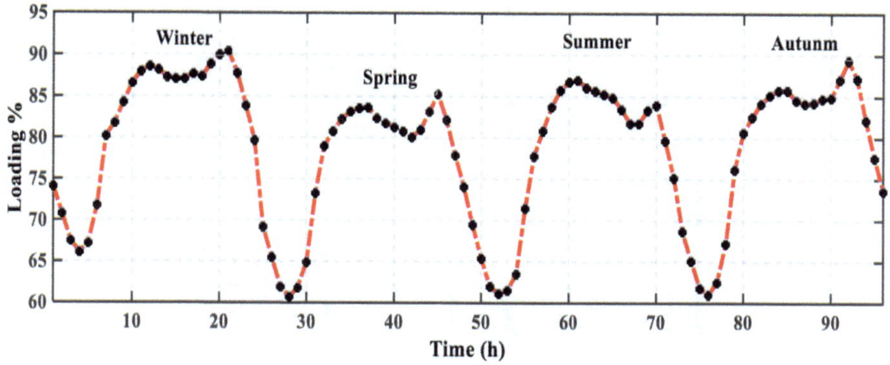


**Fig. 5.7** The seasonal hourly load profile

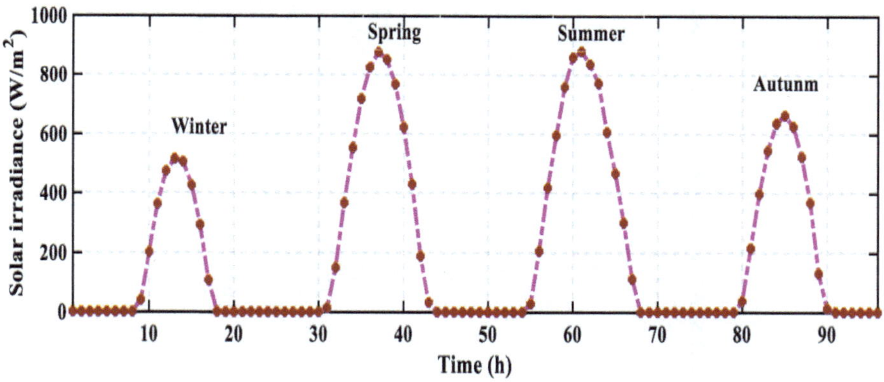


**Fig. 5.8** The seasons solar irradiance variations

**Table 5.7** The cost coefficients of the PV-DG and the D-STATCOM

| Parameter | Value |
|---|---|
| *PV cost* [162] | |
| $C_{PV}$ | 770 $/kW |
| $C_{O\ \&\ M}$ | 0.01 $/kWh |
| $\tau$ | 10% |
| NP | 20 |
| *D-STATCOM cost* [163] | |
| $C_S$ | 50 $/ kVAR |
| $\alpha$ | 10% |
| ND | 30 |
| *Grid cost* | |
| $C_{loss}$ [164] | 0.06 $/kWh |
| $C_{Grid}$ [162] | 0.096 $/kWh |

**Table 5.8** The simulation results of inclusion of single PV-DG and D-STATCOM in 69-bus system considering uncertainties of system

| Item | Base case | MALO | ALO | SCA | WOA | GOA |
|---|---|---|---|---|---|---|
| $E_{loss}$(MWh) | 1.19726 | 1.2665 | 0.6582 | 1.2646 | 1.3195 | 1.3345 |
| $E_{grid}$(MWh) | 27.44682 | 21.27 | 24.8570 | 22.9680 | 22.6080 | 26.6040 |
| Optimal location 1, Optimal location 2 | – | 58 | 62 | 58 | 62 | 57 |
| $P_{sr1}$(kW) $P_{sr2}$(kW) | – | 3389 | 1113 | 2467 | 2692 | 532 |
| $Q_{DS1}$(kVAR) $Q_{DS2}$(kVAR) | – | 2691 | 1290 | 2694 | 2586 | 2689 |
| TVD (p. u.) | 1.24867E+4 | 6.075E+3 | 8.736E+3 | 6.090E+3 | 6.127E+3 | 7.609E+3 |
| TVSI (p. u.) | 5.487872E+5 | 5.76450E+5 | 5.62160E+5 | 5.7358E+5 | 5.75870E+5 | 5.66340E+5 |
| $C_{loss}$ ($) | 7.1836199E+4 | 7.5988E+4 | 3.94910E+4 | 7.5876E+4 | 7.9171E+4 | 8.0073E+4 |
| $C_{Grid}$ ($) | 2.634895E+6 | 2.041900E+6 | 2.386200E+6 | 2.204900E+6 | 2.170400E+6 | 2.554000E+6 |
| $C_{PV}$ ($) | – | 3.68970E+5 | 1.21180E+5 | 2.68590E+5 | 2.93090E+5 | 5.7921E+4 |
| $C_{DS}$ ($) | – | 1.4273E+4 | 6.842E+3 | 1.4289E+4 | 1.3716E+4 | 1.4262E+4 |
| $C_{Total}$ ($) | 2.706731E+6 | 2.501200E+6 | 2.553700E+6 | 2.563600E+6 | 2.556300E+6 | 2.706200E+6 |

**Table 5.9** The simulation results of inclusion of two PV-DGs and D-STATCOMs in 69-bus system considering uncertainties of system

| Item | MALO | ALO | SCA | WOA | GOA |
|---|---|---|---|---|---|
| $E_{loss}$(MWh) | 0.892426 | 0.8136 | 1.3468 | 1.0280 | 1.1174 |
| $E_{grid}$(MWh) | 20.137 | 23.3281 | 21.2413 | 21.5289 | 21.6146 |
| Optimal location 1,<br>Optimal location 2 | 61<br>12 | 63<br>57 | 57<br>61 | 63<br>59 | 62<br>58 |
| $P_{sr1}$(kW)<br>$P_{sr2}$(kW) | 1646<br>2155 | 690<br>1337 | 3018<br>429 | 1868<br>1251 | 1805<br>1316 |
| $Q_{DS1}$(kVAR)<br>$Q_{DS2}$(kVAR) | 1987<br>707 | 1449<br>581 | 159<br>2503 | 1163<br>1241 | 935<br>1683 |
| TVD (p. u.) | 5.301E+3 | 6.949E+3 | 5.855E+3 | 6.333E+3 | 6.118E+3 |
| TVSI (p. u.) | 5.80198E+5 | 5.68974E+5 | 5.76828E+5 | 5.75773E+5 | 5.76570E+5 |
| $C_{loss}$ ($) | 5.3545E+4 | 4.8816E+4 | 8.0808E+4 | 6.1681E+4 | 6.70425E+4 |
| $C_{Grid}$ ($) | 1.93319E+6 | 2.23950E+6 | 2.03916E+6 | 2.06677E+6 | 2.07501E+6 |
| $C_{PV}$ ($) | 4.13822E+5 | 2.20680E+5 | 3.75311E+5 | 3.39581E+5 | 3.39798E+5 |
| $C_{DS}$ ($) | 1.4288E+4 | 1.0767E+4 | 1.4119E+4 | 1.2750E+4 | 1.3885E+4 |
| $C_{Total}$ ($) | 2.414851E+6 | 2.519764E+6 | 2.50940E+6 | 2.48078E+6 | 2.49573E+6 |

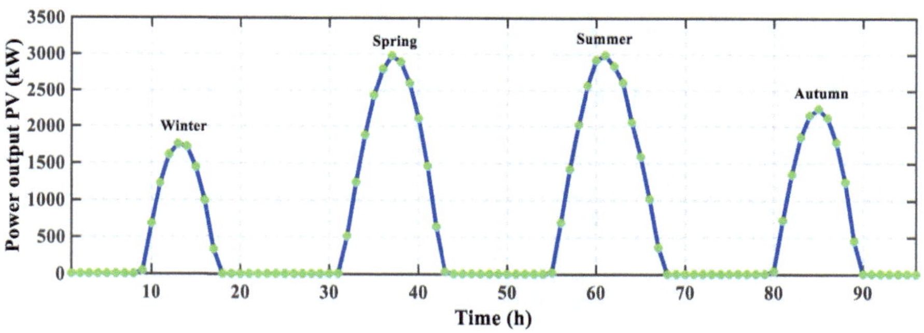


**Fig. 5.9** The hourly output power of the PV unit

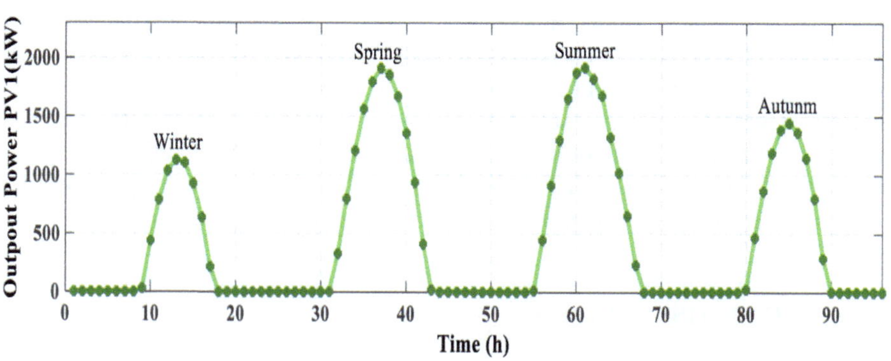


**Fig. 5.10** The output power of the first PV unit

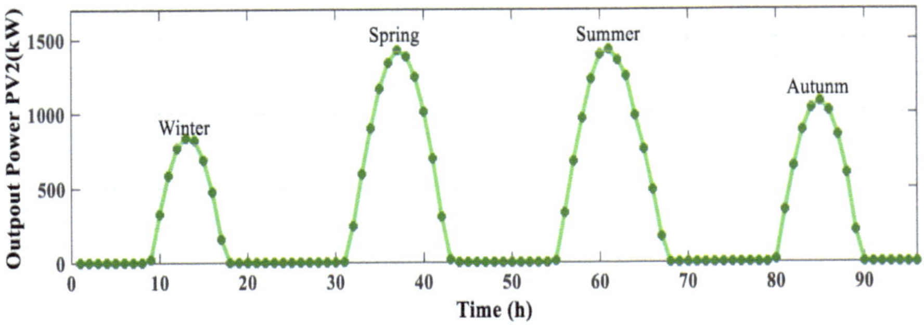


**Fig. 5.11** The output power of the second PV unit

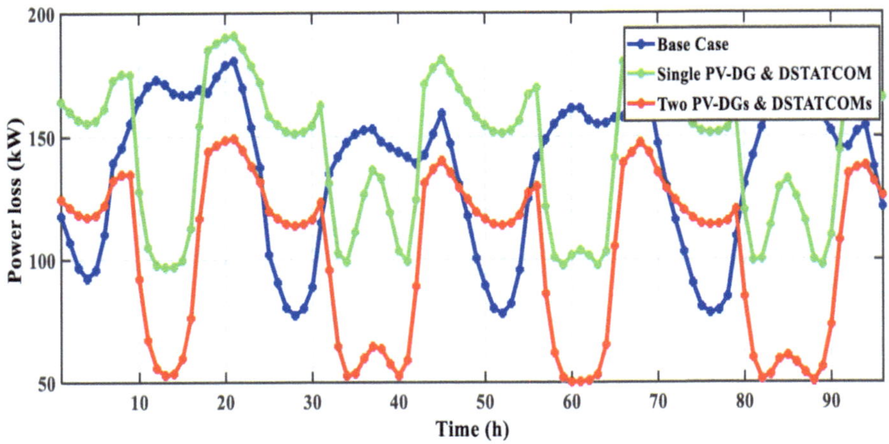


**Fig. 5.12** The power losses under uncertain conditions

Figure 5.12 shows the power losses under uncertain conditions. Figure 5.12 shows that the minimum power losses are obtained by inclusion of two hybrid systems. Figure 5.13 shows the voltage profile of the system that incorporates PV-DGs and D-STATCOMs in four seasons. According to Fig. 5.13, it is clear that the voltage profile is enhanced considerably with inclusion two hybrid systems compared with base case. Table 5.9 shows a comparison between the obtained results by the proposed algorithm, the conventional ALO, SCA, WOA, and GOA in terms of the cost, TVD, and TVSI. Judging from Table 5.9, the outcomes by application the proposed algorithm are better than the reported algorithm.

#### 5.3.2.2 IEEE-118 Bus System

In this case, the proposed algorithm is applied for solving the optimal power planning problem under uncertainty conditions for 118-bus system as a large-scale

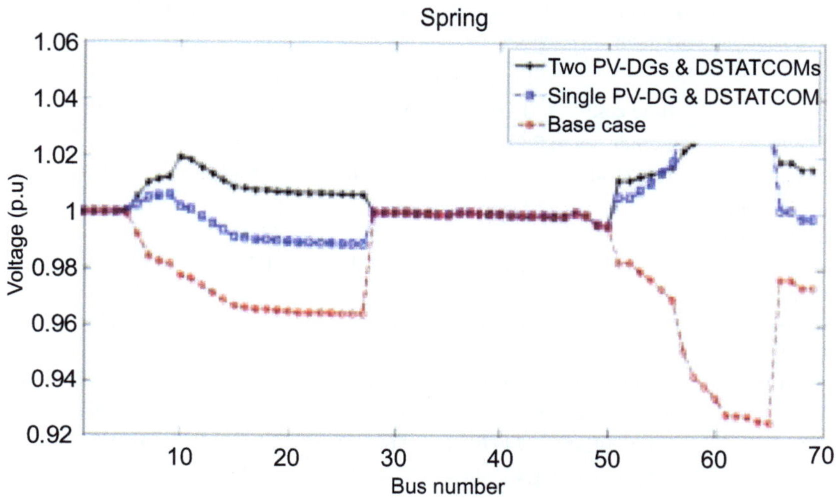


(a)

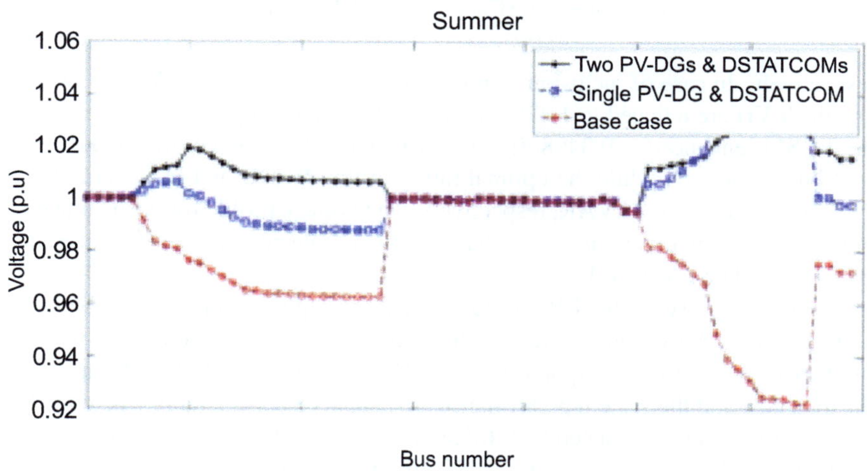


(b)

**Fig. 5.13** The voltage profile of 69-bus system by incorporating the PV-DGs and D-STATCOMs in (**a**) spring, (**b**) summer, (**c**) autumn, and (**d**) winter

system. The system load profile and solar irradiance for this case under uncertain condition are also depicted in Figs. 5.7 and 5.8, respectively. The optimal power planning problem solution is assessed with incorporating single and two hybrid systems.

The outcomes for this case are listed in Table 5.10.

At the base case (without the inclusion of PV or D-STATCOM), the total cost, **TVD**, and **TVSI** are 1.61300E+7 p.u., 3.5605E+4 p.u. and 8.93320E+5 p.u.,

**Table 5.10** The simulation results of inclusion of the PV-DGs and D-STATCOMs in 118-bus system considering uncertainties of system

| | | Single hybrid system | | Two hybrid system | |
|---|---|---|---|---|---|
| Item | Base case | ALO | MALO | ALO | MALO |
| $E_{loss}$(MWh) | 6.8999 | 7.9161 | 7.7761 | 7.7213 | 7.1755 |
| $E_{grid}$(MWh) | 163.7100 | 151.3400 | 122.7400 | 138.4801 | 122.1273 |
| Optimal location 1,<br>Optimal location 2 | – | 64 | 64 | 65<br>22 | 65<br>30 |
| $P_{sr1}$(kW)<br>$P_{sr2}$(kW) | – | 7266 | 22,709 | 10,795<br>3338 | 3498<br>18,177 |
| $Q_{DS1}$(kVAR)<br>$Q_{DS2}$(kVAR) | – | 16,030 | 14,642 | 10,072<br>2300 | 12,676<br>9308 |
| TVD (p. u.) | 3.5605E+4 | 2.5614E+4 | 2.5869E+4 | 2.4357E+4 | 2.1244E+4 |
| TVSI (p. u.) | 8.93320E+5 | 9.33530E+5 | 9.34980E+5 | 9.40625E+5 | 9.51969E+5 |
| $C_{loss}$ ($) | 4.13990E+5 | 4.74970E+5 | 4.66570E+5 | 4.63276E+5 | 4.30531E+5 |
| $C_{Grid}$ ($) | 1.57160E+7 | 1.45280E+7 | 1.17830E+7 | 1.32941E+7 | 1.17242E+7 |
| $C_{PV}$ ($) | – | 7.9107E+5 | 2.4724E+6 | 1.5388E+6 | 2.4722E+6 |
| $C_{DS}$ ($) | – | 8.5023E+4 | 7.7661E+4 | 5.3421E+4 | 2.7013E+4 |
| $C_{Total}$ ($) | 1.61300E+7 | 1.58790E+7 | 1.47990E+7 | 1.53496E+7 | 1.465397E+7 |

respectively. In case of inclusion a single PV-DG and D-STATCOM, the total cost and the **TVD** are reduced to 1.47990E+7 $ and 2.5869E+4 p.u., respectively, while the **TVSI** is enhanced to 9.34980E+5 p.u. The optimal location of the hybrid system is at bus number 64 while the optimal rating of the PV-DG and D-STATCOM are 22,709 kW and 14,642 kVAR, respectively. The voltage profile for this is illustrated in Fig. 5.13. From this figure, the voltage profile is enhanced with inclusion a single PV-DG and D-STATCOM.

The output power of the PV unit for this case is depicted in Fig. 5.14. According to this figure, the output power is changed hourly with variations of the solar irradiance. In case of incorporating two PV-DGs and D-STATCOMs in system, the total cost and the system performance are enhanced considerably where the total cost and the TVD are reduced to 1.465397E+7 $ and 2.1244E+4 p.u., respectively, while the TVSI is enhanced to 9.51969E+5 p.u. in other word the obtained (Figs. 5.15, 5.16, and 5.17).

## 5.4 Optimal Allocation of DER in RDS Using LAPO and EO Compared with WOA and SCA

### *5.4.1 IEEE-118 Bus Radial Distribution System*

In this section, the proposed method, including the sensitivity analysis and the selected optimization algorithm, is applied to determine the optimal sizes and locations for the PV unit and D-STATCOM. The proposed method is applied to

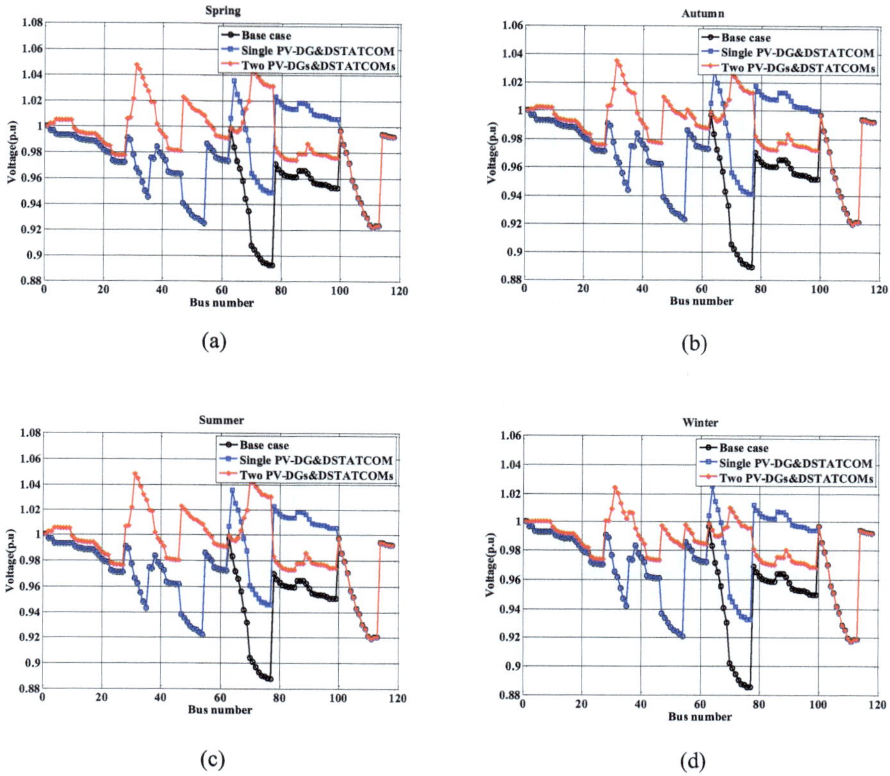


**Fig. 5.14** The voltage profile of 118-bus system by incorporating the PV-DGs and D-STATCOMs in (**a**) spring, (**b**) summer, (**c**) autumn, and (**d**) winter

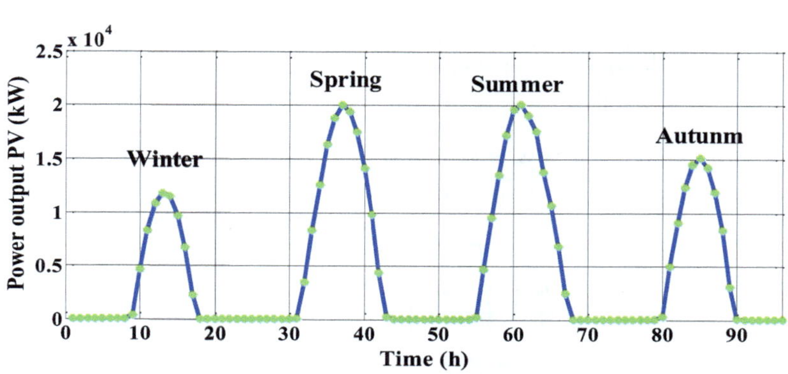


**Fig. 5.15** The hourly output power of the single PV unit in 118-bus system

the IEEE-118 bus distribution as a large-scale system. The single line diagram of the system is shown in Fig. 5.18, while the system specifications are listed in Table 5.11 and the line and bus data are given in [153]. The program code was written using MATLAB software and the simulations were carried out on a 2.5 GHz core I5 with

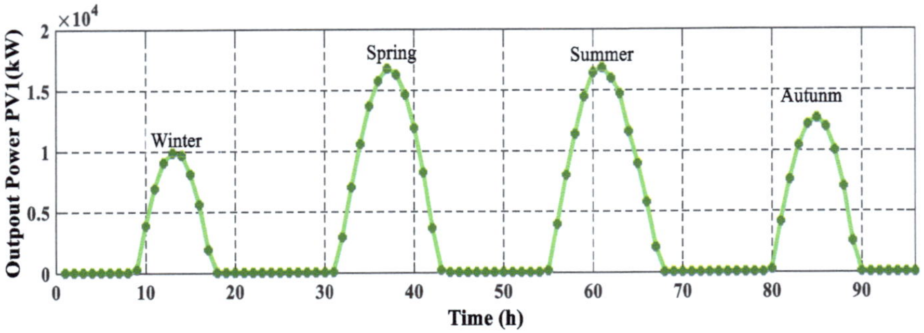


**Fig. 5.16** The output power of the first PV unit in the 118-bus system

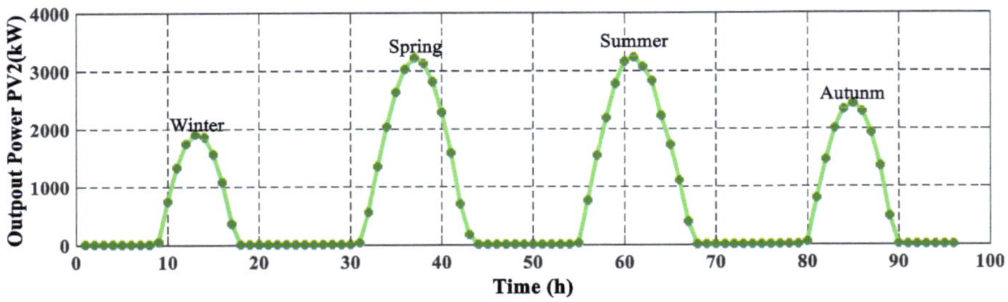


**Fig. 5.17** The output power of the second PV unit in the 118-bus system

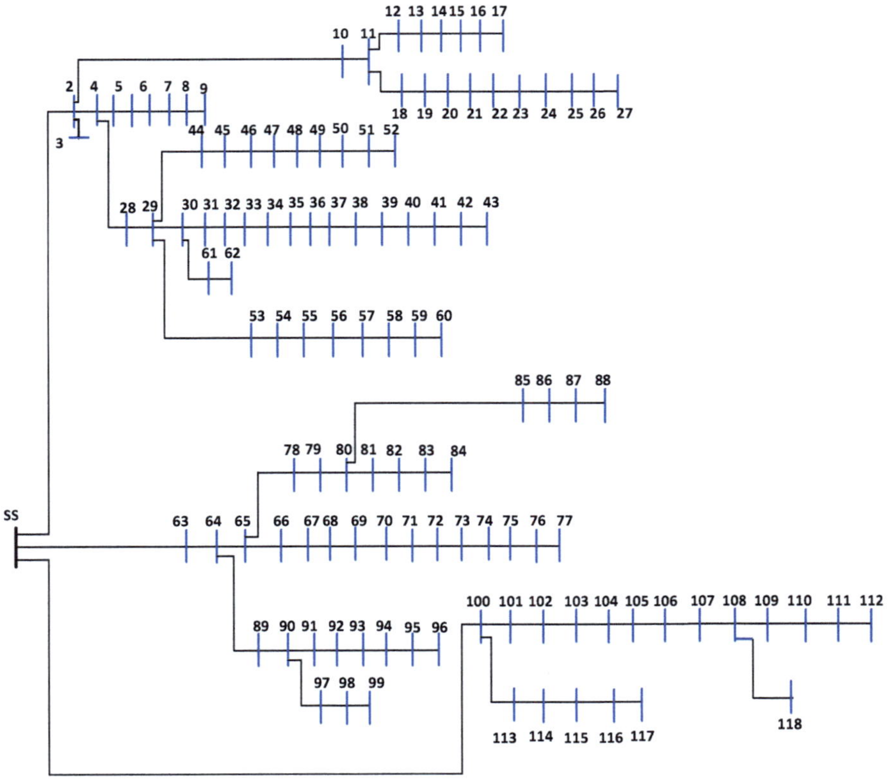


**Fig. 5.18** Single line diagram of 118-bus system

**Table 5.11** The specification of the 118-bus system

| Item | Value |
|---|---|
| *System specifications:* | |
| NB | 118 |
| Nt | 117 |
| System voltage (kV) | 11 |
| Base MVA | 100 |
| $S_{\text{Load}}$(MVA) | 22.709 + j17.041 |
| *Initial power flow results:* | |
| $P_{\text{Total}_{\text{loss}}}$ (kW) | 1298.091 |
| $Q_{\text{Total}_{\text{loss}}}$ (kVar) | 978.797 |
| $V_{\min}$ (p. u.) @ bus | 0.86880 @ 77 |

**Table 5.12** The selected parameters of the optimization algorithms

| Algorithm | Parameter settings |
|---|---|
| LAPO | $T_{\max} = 100$, Search agents No. = 25 |
| EO | $T_{\max} = 100$, Search agents No. = 25, $a1 = 2$, $a2 = 1$, GP = 0.5 |
| WOA | $T_{\max} = 100$, Search agents No. = 25 |
| SCA | $T_{\max} = 100$, Search agents No. = 25 |

4 GB of PC. Two studied cases are presented, including optimal integration of a single PV unit with a single D-STATCOM, while in the second case, two PV units with two D-STATCOMs are incorporated. Lines and bus data for this system are given in Appendix D. The investigation is performed under three perpetration levels of the PV-based DG, including 20%, 30%, and 40%.

The cost coefficients of the PV, D-STATCOM, and grid are listed in Table 5.7. Two recent optimization techniques are applied to solve the allocation problem, including LAPO and EO techniques, and the obtained results are compared with WOA and SCA techniques. The selected parameters of these algorithms are listed in Table 5.12, and for fair comparison, the maximum number of iterations and search agents are selected the same. A sensitivity analysis is performed to determine the candidate buses for integration of PV and D-STATCOM based on the voltage loss sensitivity index VLSI as illustrated in Fig. 5.19 shows the VLSI of system for each bus. The highest candidate buses are 70, 64, 104, 2, 68, 106, 101, 103, 4, 31, 69, 67, 78, 30, 108, 102, 66, 105, 100, 89, 63, 107, 33, 32,34, 110, 47, 28, 35, 109, 38, 71, 49, 79, 72, 39, 50, 10, 29, 48, 73, 42, 90, 91, 18, 74, 11, 111, 55, 58, 41, 80, 56, 40, 23, 20, 96, 19, and 57.

Practically, the load profile and the solar irradiance are varied during the day and seasonally in spring, summer, autumn, and winter. Thus, the investigation is performed while considering these variations. The hourly load profile is depicted in Fig. 5.20 [143], and the solar irradiance is depicted in Fig. 5.21. Figure 5.22 shows the power losses in the system of four seasons in the base case (without PV or D-STATCOM), while Fig. 5.23 shows the drawn power from the grid. Referring to

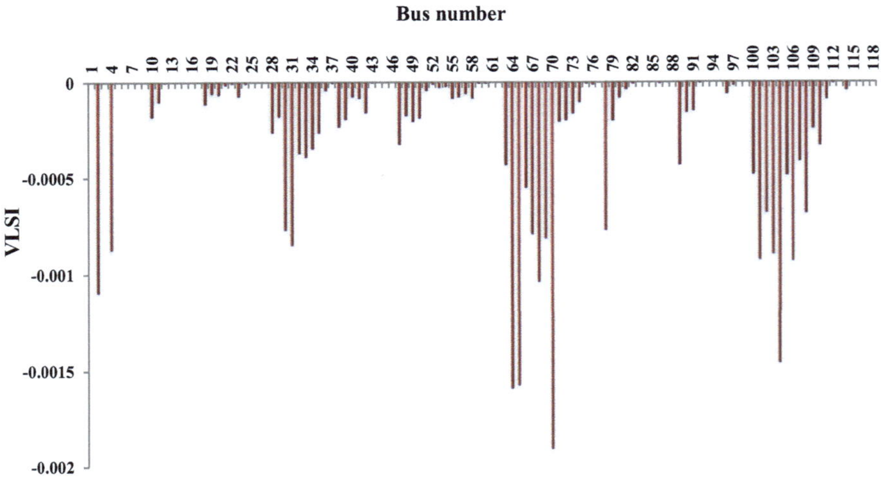


**Fig. 5.19** The VLSI for 118-bus system

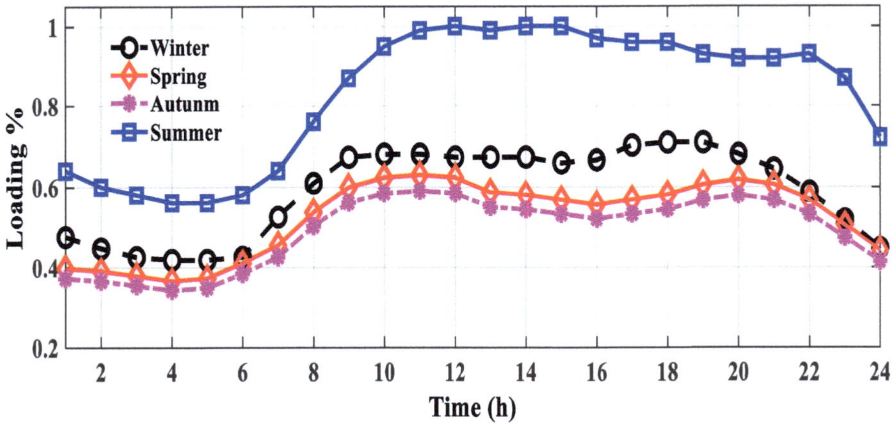


**Fig. 5.20** The hourly load profile of four seasons

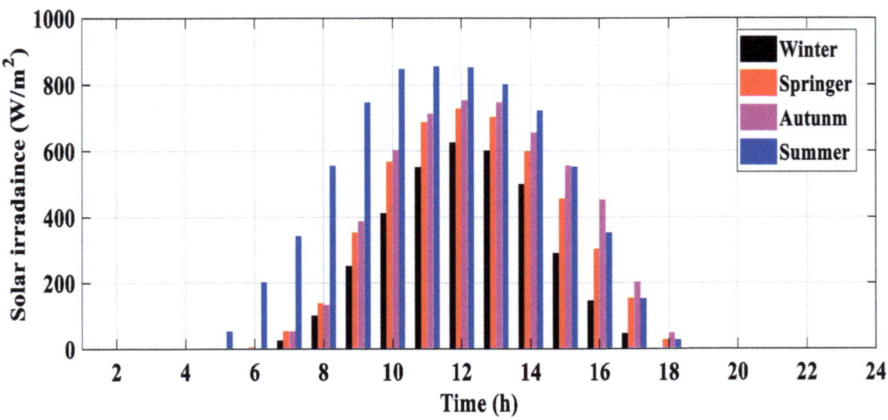


**Fig. 5.21** The hourly solar irradiance of four seasons

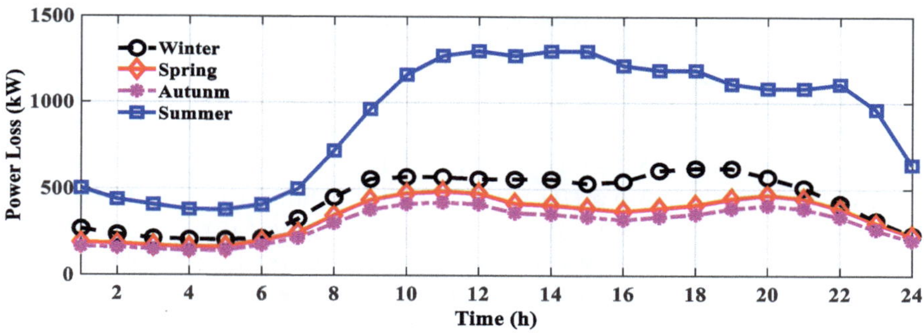


**Fig. 5.22** The power losses of four seasons at base case

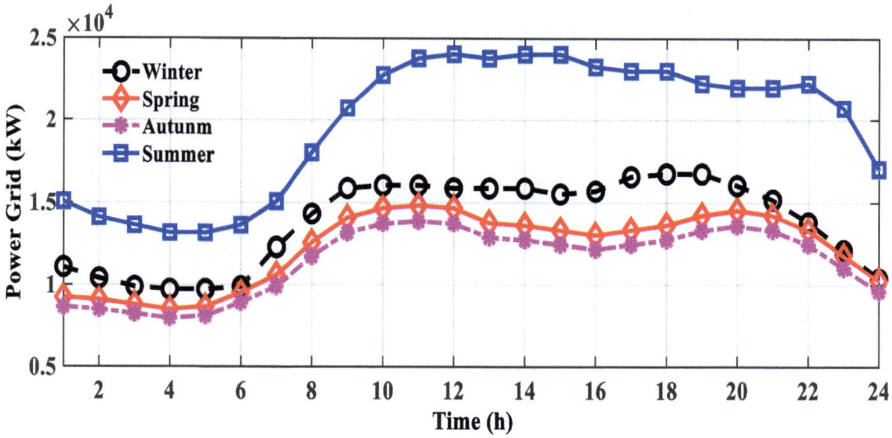


**Fig. 5.23** The purchased power from grid at base case

Fig. 5.22, the power losses vary with variations in seasonal load demands where the highest power losses occur in the summer. Also, referring to Fig. 5.23, the drawn powers from the grid increase or decrease with variations in the load demand; the highest power drawn occurred in summer. The studied cases are represented as follows:

#### 5.4.1.1 Case 1: Single PV and Single (D-STATCOM)

In this case, the proposed method is applied to assigning the optimal sizes and locations of a single PV and a single D-STATCOM using EO, LAPO, WOA, and SCA. Without the inclusion of PV and D-STATCOM, the annul energy loss, the purchased energy from the grid, and the total cost are 4.332800E+06, 1.252401E +08, and 1.2283024E+07, respectively. The investigation was performed under different values of penetration levels. The obtained results under 20%, 30%, and 40% penetration levels are tabulated in Tables 5.13, 5.14, and 5.15, respectively.

Table 5.13 A comparison of the investigated algorithms for case 1 under 20% penetration level

| Item | EO | LAPO | WOA | SCA |
|---|---|---|---|---|
| $E_{loss}$(kWh) | 3.62554E+06 | 3.62550E+06 | 3.64912E+06 | 3.65478E+06 |
| $E_{grid}$(kWh) | 1.16099E+08 | 1.16153E+08 | 1.177999E+08 | 1.16233E+08 |
| $P_{sr}$(kW) @ bus | 4541@ (71) | 4512 @ (71) | 3638 @ (70) | 4485 @ (70) |
| $Q_{DS}$(kVAR) @ bus | 1380 @ (71) | 1397 @ (71) | 1675@ (70) | 1458@ (70) |
| TVD (p. u.) | 2.7163E+04 | 2.2588E+04 | 2.2566E+04 | 2.2281E+04 |
| TVSI (p. u.) | 9.23084E+05 | 9.41777E+05 | 9.41827E+05 | 9.41498E+05 |
| $Cost_{loss}$ ($) | 2.17532E+05 | 2.17530E+05 | 2.18947E+05 | 2.19286E+05 |
| $Cost_{Grid}$ ($) | 1.11455E+07 | 1.11507E+07 | 1.13088E+07 | 1.11583E+07 |
| $Cost_{PV}$ ($) | 4.95041E+05 | 4.91880E+05 | 3.96600E+05 | 4.88936E+05 |
| $Cost_{DS}$ ($) | 7.319E+03 | 7.409E+03 | 8.884E+03 | 7.733E+03 |
| $Cost_{Total}$ ($) | 1.18654E+07 | 1.18675E+07 | 1.19332E+07 | 1.18743E+07 |
| Reduction in cost (%) | 3.4 | 3.38 | 2.85 | 3.33 |

Table 5.14 A comparison of the investigated algorithms for case 1 under 30% penetration level

| Item | EO | LAPO | WOA | SCA |
|---|---|---|---|---|
| $E_{loss}$(kWh) | 3.79140E+06 | 3.80030E+06 | 4.124487E+06 | 3.90053E+06 |
| $E_{grid}$(kWh) | 1.120474E+08 | 1.12110E+08 | 1.12380E+08 | 1.12248E+08 |
| $P_{sr}$(kW) @ bus | 6812 @ (67) | 6783@ (66) | 6737@ (64) | 6763 @ (65) |
| $Q_{DS}$(kVAR) @ bus | 1409 @ (67) | 1514 @ (66) | 1434 @ (64) | 1909@ (65) |
| TVD (p. u.) | 2.2621E+04 | 2.2820E+04 | 2.2441E+04 | 9.39875E+05 |
| TVSI (p. u.) | 9.39441E+05 | 9.38412E+05 | 9.39659E+05 | 9.39875E+05 |
| $Cost_{loss}$ ($) | 2.27483E+05 | 2.28018E+05 | 2.47469E+05 | 2.34031E+05 |
| $Cost_{Grid}$ ($) | 1.07566E+07 | 1.07626E+07 | 1.07885E+07 | 1.07758E+07 |
| $Cost_{PV}$ ($) | 7.42617E+05 | 7.39455E+05 | 7.42617E+05 | 7.37275E+05 |
| $Cost_{DS}$ ($) | 7.473E+03 | 8.030E+03 | 1.2045E+04 | 1.0125E+04 |
| $Cost_{Total}$ ($) | 1.17341E+07 | 1.17381E+07 | 1.17907E+07 | 1.17572E+07 |
| Reduction in cost (%) | 4.47 | 4.44 | 4.01 | 4.28 |

Referring to Table 5.13, at 20% of penetration level, the total costs obtained by EO, LAPO, WOA, and SCA are 1.186542E+07$, 1.186752E+07$, 1.193322E+07$, and 1.187428E+07$, respectively. Thus, the minimum value of the total cost is obtained by the application of the EO technique. In this case, the purchased energy from the grid is reduced from 1.252401E+08 kWh to 1.16099E+08 kWh. In this case, the optimal allocation of PV units and the D-STATCOM is 4541 kW at bus 71 and 1380 kVAR at bus 71, respectively. The total costs obtained by EO, LAPO, WOA, and SCA at 30% penetration level are 1.1734128E+07$, 1.1738083E+07$, 1.1790662E+07$, and 1.1757199E+07$, respectively, according to Table 5.14.

The best results can be obtained by application the EO technique. At this case, the purchased energy from grid is reduced from 1.252401E+08 kWh to 1.120474E+08 kWh. The optimal ratings and sizes and PV unit and DSATACOM are 6812 kW at bus 67 and 1409 kVAR at bus 67, respectively. Also, from

**Table 5.15** A comparison of the investigated algorithms for case 1 under 40% penetration level

| Item | EO | LAPO | WOA | SCA |
|---|---|---|---|---|
| $E_{loss}$(kWh) | 3.81649E+06 | 3.83579E +06 | 4.16819E+06 | 4.01644E+06 |
| $E_{grid}$(kWh) | 1.07855E+08 | 1.08273E+08 | 1.08208E+08 | 1.08403E+08 |
| $P_{sr}$(kW) @ bus | 9083@ (65) | 8868@ (66) | 9082@ (64) | 8895@ (102) |
| $Q_{DS}$(kVAR) @ bus | 1411@ (65) | 1443 @ (66) | 2375@ (64) | 1921@ (102) |
| **TVD(p. u.)** | 2.2865E+04 | 2.2551E+04 | 2.2083E+04 | 2.29411E+04 |
| **TVSI(p. u.)** | 9.38300E+05 | 9.39643E+05 | 9.40985E+05 | 9.37850E+05 |
| $Cost_{loss}$ ($) | 2.28989E+05 | 2.30147E+05 | 2.50091E+05 | 2.40986E+05 |
| $Cost_{Grid}$ ($) | 1.03541E+07 | 1.03942E+07 | 1.03880E+07 | 1.04067E+07 |
| $Cost_{PV}$ ($) | 9.90192E+05 | 9.66754E+05 | 9.90083E+05 | 9.69697E+05 |
| $Cost_{DS}$ ($) | 7.489E+03 | 7.653E+03 | 1.2596E+04 | 1.01889E+04 |
| $Cost_{Total}$ ($) | 1.1580731E+07 | 1.1598800E+07 | 1.1640775E+07 | 1.16276E+07 |
| Reduction in cost (%) | 5.72 | 5.57 | 5.23 | 5.34 |

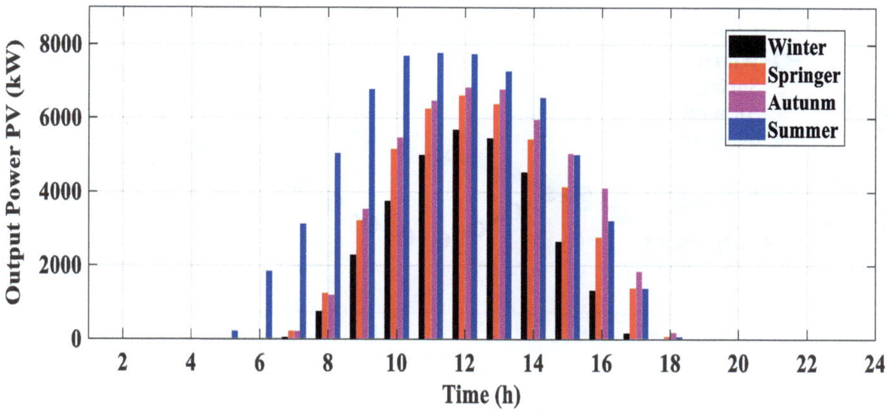


**Fig. 5.24** The hourly output power of PV unit

Table 5.15, the minimum total cost at 40% penetration level is 1.1580731E+07$ which obtained by EO. At this case, the purchased energy from grid is reduced from 1.252401E+08 kWh to 1.07855E+08 kWh. The optimal ratings and placement for of PV unit and DSATACOM for this case are 9083 kW at bus 65 and 1411 kVAR at bus 65. Judging from Tables 5.13, 5.14, and 5.15, it is obvious that the total cost decreases with increasing of the penetration levels of the injected PV power. Figure 5.24 shows the output power of the PV unit the output PV power follows the variations of hourly solar irradiance consequently, the power loss and the drawn power decrease considerably as depicted in Figs. 5.25 and 5.26 especially at high values of the solar irradiance compared with the base case.

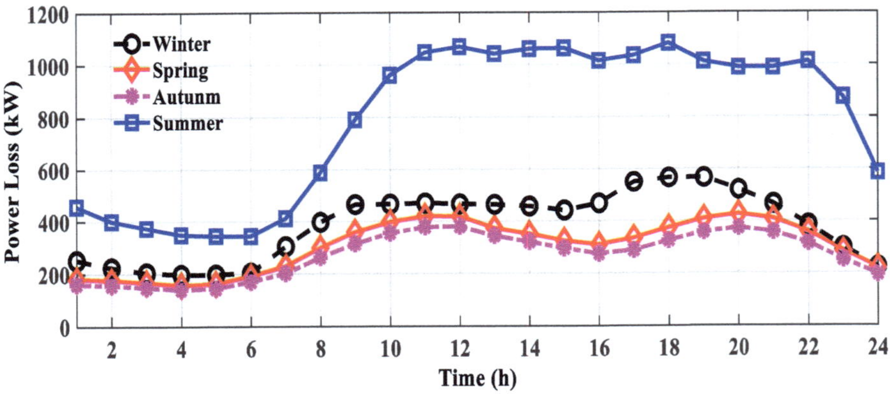


**Fig. 5.25** The power losses of four seasons for case 1

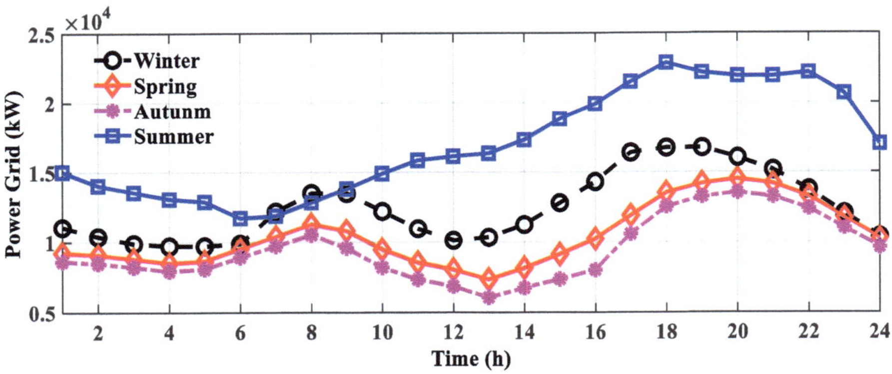


**Fig. 5.26** The purchased power for case 1

#### 5.4.1.2 Case 2: Two PV Units and Two (D-STATCOMs)

In this case, the proposed method is applied to find the optimal allocation of two PV units and two D-STATCOMs. Tables 5.16, 5.17, and 5.18 show the obtained results at 20%, 30%, and 40% penetration levels, respectively. The optimal costs at 20%, 30%, and 40% penetration levels are 1.1776987E+07 $, 1.1601088E+07 $, 1.216021E+07 $, and 1.146072E+07 $, respectively. Judging from Tables 5.16, 5.17, and 5.18, the inclusion of two PV units and two D-STATCOMs can minimize the total cost compared with case 1 from 1.1865427E+07 $ to 1.1776987E+07$, from 1.1734128E+07$ to 1.1601088E+07$ and from 1.1580731E+07$ to 1.146072E+07$ at 20%, 30% and 40% penetration levels, respectively. At the 40% penetration level, the optimal allocations of the PV units are 5884 kW at bus 107 and 3199 kW at bus 70, while the optimal allocations of D-STATCOMs are

**Table 5.16** A comparison of the investigated algorithms for case 2 under 20% penetration level

| Item | EO | LAPO | WOA | SCA |
|---|---|---|---|---|
| $E_{loss}$(kWh) | 2.99661E+06 | 3.23532E+06 | 3.87473E+06 | 3.38373E+06 |
| $E_{grid}$(kWh) | 1.15469E+08 | 1.16615E+08 | 1.23087E+08 | 1.15931E+08 |
| $P_{sr1}$(kW) @ bus<br>$P_{sr2}$(kW) @ bus | 2519 @ (111)<br>2022 @ (70) | 2421 @ (70)<br>1631@ (110) | 404@ (104)<br>509@ (105) | 4339 @ (70)<br>157 @ (89) |
| $Q_{DS1}$(kVAR) @ bus<br>$Q_{DS2}$(kVAR) @ bus | 1418@ (70)<br>1805 @ (110) | 1450 @ (70)<br>807 @ (104) | 407 @ (104)<br>1831 @ (70) | 969@ (48)<br>742 @ (70) |
| **TVD (p. u.)** | 2.0923E+04 | 2.1965E+04 | 2.3332E+04 | 2.1994E+04 |
| **TVSI (p. u.)** | 9.45278E+05 | 9.41312E+05 | 9.36391E+05 | 9.42314E+05 |
| $Cost_{loss}$ ($) | 1.79796E+05 | 1.94118E+05 | 2.32483E+05 | 2.03023E+05 |
| $Cost_{Grid}$ ($) | 1.10850E+07 | 1.11951E+07 | 1.18163E+07 | 1.11294E+07 |
| $Cost_{PV}$ ($) | 4.95053E+05 | 4.41751E+05 | 9.9529E+04 | 4.90235E+05 |
| $Cost_{DS}$ ($) | 1.7094E+04 | 1.1971E+04 | 1.1870E+04 | 9.0750E+03 |
| $Cost_{Total}$ ($) | 1.17769E+07 | 1.18429E+07 | 1.21602E+07 | 1.18317E+07 |
| Reduction in cost (%) | 4.12 | 3.58 | 1.00 | 3.67 |

**Table 5.17** A comparison of the investigated algorithms for case 2 under 30% penetration level

| Item | EO | LAPO | WOA | SCA |
|---|---|---|---|---|
| $E_{loss}$(kWh) | 2.89237E+06 | 3.3603E+06 | 4.06896E+06 | 3.43954E+06 |
| $E_{grid}$(kWh) | 1.11147E+08 | 1.11705E+08 | 1.123412E+08 | 1.12029E+08 |
| $P_{sr1}$(kW) @ bus<br>$P_{sr2}$(kW) @ bus | 3707@ (109)<br>3105@ (70) | 2298 @ (110)<br>4469@ (70) | 386 @ (70)<br>415@ (110) | 3423 @ (70)<br>3209 @ (31) |
| $Q_{DS1}$(kVAR) @ bus<br>$Q_{DS2}$(kVAR) @ bus | 1455 @ (70)<br>1336@ (110) | 100 @ (64)<br>1568 @ (70) | 37 @ (104)<br>1716@ (71) | 1238 @ (70)<br>1091 @ (65) |
| **TVD (p. u.)** | 2.0519E+04 | 2.1450E+04 | 2.2838E+04 | 2.1639E+04 |
| **TVSI (p. u.)** | 9.47528E+05 | 9.45669E+05 | 9.403453E+05 | 9.43388E+05 |
| $Cost_{loss}$ ($) | 1.73542E+05 | 2.01618E+05 | 2.44137E+05 | 2.06372E+05 |
| $Cost_{Grid}$ ($) | 1.06701E+07 | 1.16718E+07 | 1.07848E+07 | 1.07548E+07 |
| $Cost_{PV}$ ($) | 7.42631E+05 | 7.37659E+05 | 7.42455E+05 | 7.22999E+05 |
| $Cost_{DS}$ ($) | 1.4803E+04 | 8.847E+03 | 9.451E+03 | 1.2352E+04 |
| $Cost_{Total}$ ($) | 1.16011E+07 | 1.16719E+07 | 1.17808E+07 | 1.16965E+07 |
| Reduction in cost (%) | 5.55 | 4.98 | 4.09 | 4.78 |

1567 kVAR at bus 70 and 1056 kVAR at bus 110. Referring to Table 5.16 at 20% of the penetration level, the purchased energy from the grid is reduced from 1.252401E+08 kWh to 1.15469E+08 kWh. Referring to Table 5.17, at 30% of the penetration level, the purchased energy from the grid is reduced from 1.252401E+08 kWh to 1.11147E+08 kWh. Referring to Table 5.13, at 40% of the penetration level, the purchased energy from the grid is reduced from 1.252401E+08 kWh to 1.07041E+08 kWh. Figure 5.27 shows the total cost for cases 1 and 2 at different penetration levels. From this figure, it is obvious that the total cost is reduced by increasing the

**Table 5.18** A comparison of the investigated algorithms for case 2 under 40% penetration level

| Item | EO | LAPO | WOA | SCA |
|---|---|---|---|---|
| **$E_{loss}$(kWh)** | 3.00953E+06 | 3.38729E+06 | 3.35108E+06 | 3.57472E+06 |
| **$E_{grid}$(kWh)** | 1.07041E+08 | 1.07426E+08 | 1.07773E+08 | 1.08970E+08 |
| **$P_{sr1}$(kW)** @ bus<br>**$P_{sr2}$(kW)** @ bus | 5884 @ (107)<br>3199 @ (70) | 4415@ (70)<br>4668@ (108) | 4584 @ (19)<br>4292 @ (70) | 4257 @ (65)<br>4095@ (71) |
| **$Q_{DS1}$(kVAR)** @ bus<br>**$Q_{DS2}$(kVAR)** @ bus | 1567 @ (70)<br>1056 @ (110) | 1812 @ (70)<br>100 @ (57) | 1813@ (70)<br>1168 @ (50) | 1362 @ (70)<br>409 @ (64) |
| **TVD (p. u.)** | 2.0417E+04 | 2.0607E+04 | 1.9474E+04 | 2.2015E+04 |
| **TVSI (p. u.)** | 9.49096E+05 | 9.49423E+05 | 9.53321E+05 | 9.44644E+05 |
| **$Cost_{loss}$** ($) | 1.80571E+05 | 2.03237E+05 | 2.01065E+05 | 2.14483E+05 |
| **$Cost_{Grid}$** ($) | 1.027598E+07 | 1.03129E+07 | 1.03462E+07 | 1.04611E+07 |
| **$Cost_{PV}$** ($) | 9.90256E+05 | 9.90186E+05 | 9.67633E+05 | 9.10505E+05 |
| **$Cost_{DS}$** ($) | 1.3912E+04 | 1.0141E+04 | 1.5811E+04 | 9.393E+03 |
| **$Cost_{Total}$** ($) | 1.14607E+07 | 1.151648E+07 | 1.153073E+07 | 1.159553E+07 |
| Reduction in cost (%) | 6.69 | 6.24 | 6.12 | 5.60 |

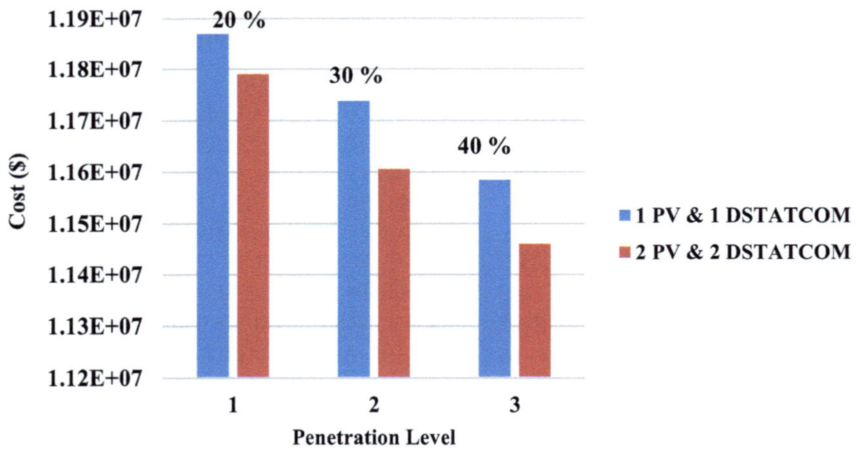


**Fig. 5.27** The minimum total costs at different penetration levels

penetration levels of the injected power by PV, as the cost decreases considerably with the inclusion of 2 PV units and 2 D-STATCOMs.

Figures 5.28 and 5.29 show the output powers of the first and second PV units, respectively. From these figures, the output powers of the PV units follow the hourly solar irradiance where the significant variations start from 8:00 AM to 17:00 PM. Figure 5.30 shows the system losses, while Fig. 5.31 shows the drawn power from the grid. From these figures, it is obvious that the system losses and drawn power decrease considerably with the injected power from the PV units. Table 5.19 shows the statistical results for case 1 with the inclusion of a single PV and a single D-STATCOM that have been obtained by the applications EO, LAPO, WOA, and

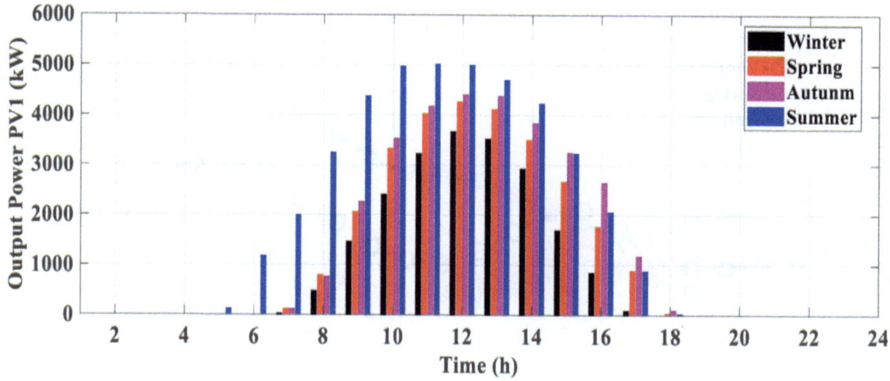


Fig. 5.28 The output power of the first PV unit

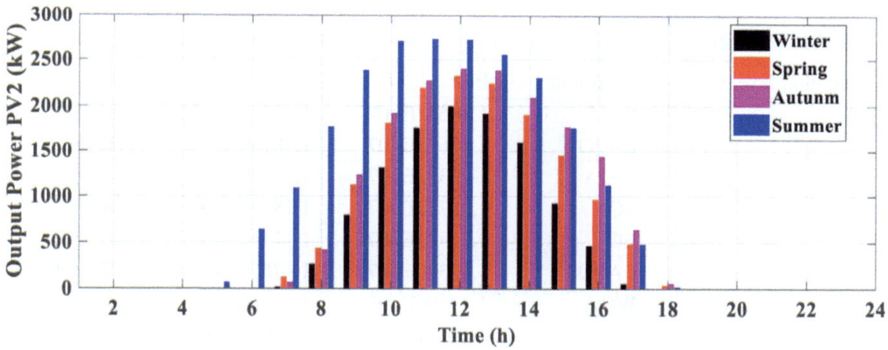


Fig. 5.29 The output power of the second PV unit

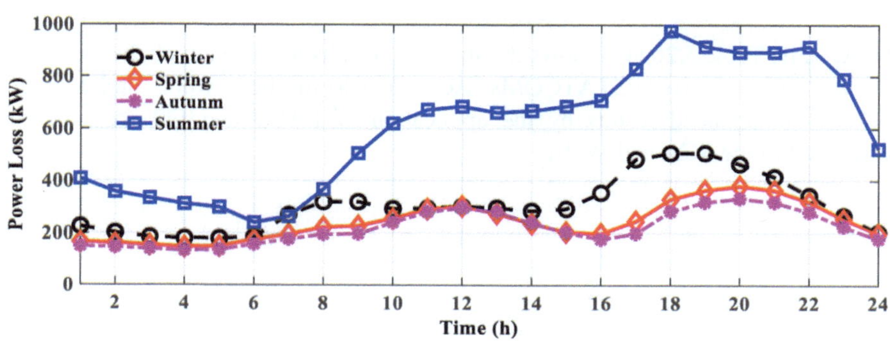


Fig. 5.30 The power losses of four seasons for case 2

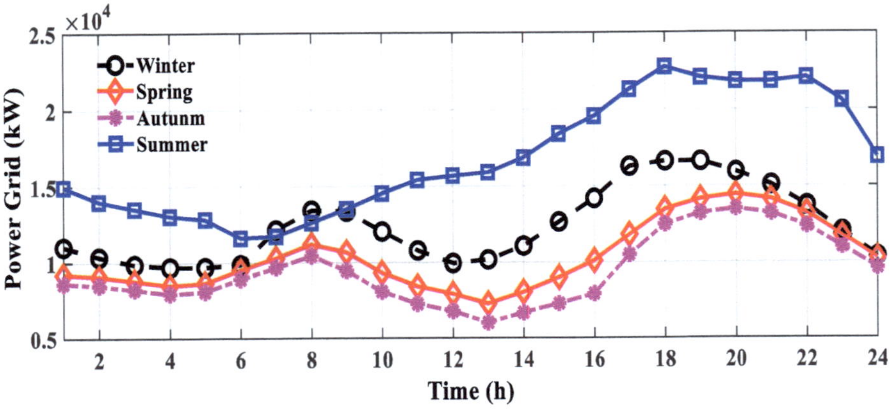


**Fig. 5.31** The purchased power from grid for case 2

**Table 5.19** Statistical results for case 1 with inclusion single PV and single D-STATCOM

| Penetration level (%) | Algorithm | Best | Average | Worst |
|---|---|---|---|---|
| 20 | EO | 1.18654E+07 | 1.18655E+07 | 1.18657E+07 |
| | LAPO | 1.18675E+07 | 1.18750E+07 | 1.19186E+07 |
| | WOA | 1.19332E+07 | 1.19400E+07 | 1.19444E+07 |
| | SCA | 1.18742E+07 | 1.18750E+07 | 1.18798E+07 |
| 30 | EO | 1.17341E+07 | 1.17344E+07 | 1.17351E+07 |
| | LAPO | 1.17380E+07 | 1.17670e+07 | 1.17898E+07 |
| | WOA | 1.17906E+07 | 1.18250E+07 | 1.19335E+07 |
| | SCA | 1.17571E+07 | 1.17750E+07 | 1.17910E+07 |
| 40 | EO | 1.15807E+07 | 1.15820E+07 | 1.15842E+07 |
| | LAPO | 1.15988E+07 | 1.15992E+07 | 1.15996E+07 |
| | WOA | 1.16407E+07 | 1.16408E+07 | 1.16409E+07 |
| | SCA | 1.16276E+07 | 1.16460E+07 | 1.16710E+07 |

SCA, while Table 5.20 also shows the statistical results for case 2 with the inclusion of two PVs and two D-STATCOMs. According to the best, worst, and average values, the obtained results by the application of LAPO and EO are superior compared to the SCA and WOA.

**Table 5.20** Statistical results for case 2 with inclusion two PV units and two D-STATCOMs

| Penetration level (%) | Algorithm | Best | Average | Worst |
|---|---|---|---|---|
| 20 | EO | 1.17769E+07 | 1.17800E+07 | 1.18128E+07 |
| | LAPO | 1.18429E+07 | 1.18477E+07 | 1.18525E+07 |
| | WOA | 1.21602E+07 | 1.21841E+07 | 1.22080E+07 |
| | SCA | 1.18317E+07 | 1.18400E+07 | 1.18586E+07 |
| 30 | EO | 1.16010E+07 | 1.16250E+07 | 1.16643E+07 |
| | LAPO | 1.16718E+07 | 1.16806E+07 | 1.16894E+07 |
| | WOA | 1.17808E+07 | 1.18750E+07 | 1.21655E+07 |
| | SCA | 1.16965E+07 | 1.185000E+07 | 1.22478E+07 |
| 40 | EO | 1.1460E+07 | 1.14632E+07 | 1.14657E+07 |
| | LAPO | 1.15164E+07 | 1.15515E+07 | 1.15867E+07 |
| | WOA | 1.15307E+07 | 1.15500E0 + 07 | 1.15825E+07 |
| | SCA | 1.15955E+07 | 1.16000E+07 | 1.16307E+07 |

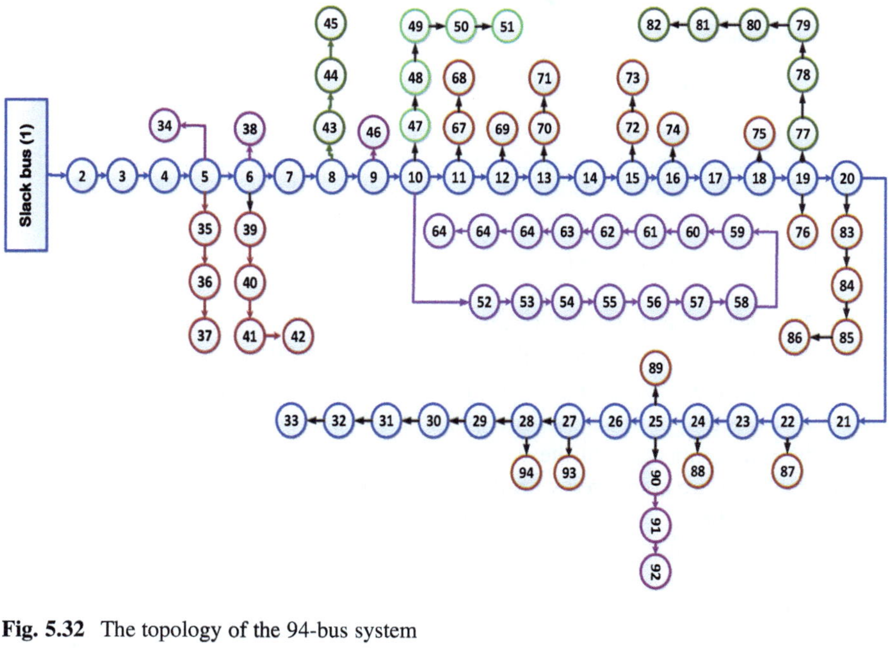


**Fig. 5.32** The topology of the 94-bus system

## 5.5 Optimal Allocation of DER in RDS Using MPA and PSO

### *5.5.1 IEEE-94 Bus Radial Distribution System*

The optimal integration of the hybrid PV-DG and D-STATCOM using the MPA has been assessed on the 94-bus network. MATLAB (2018b) software is used for writing the code in a Core I7 PC 2.50 GHz with 4 GB RAM. The single line diagram of the network is depicted in Fig. 5.32 The specifications of system are depicted in

**Table 5.21** the specification of the 94-bus system

| Item | Value |
|---|---|
| *System specifications:* | |
| NB | 94 |
| NT | 93 |
| Vsys (kV) | 15 |
| Base MVA | 100 |
| $S_{Load}$(MVA) | 4797+ j2323.900 |
| $P_{Total_{loss}}$ (kW) | 365.173 |
| $Q_{Total_{loss}}$ (kVar) | 505.785 |
| $V_{min}$ (p. u.) @ bus | 0.84749 @ bus 92 |

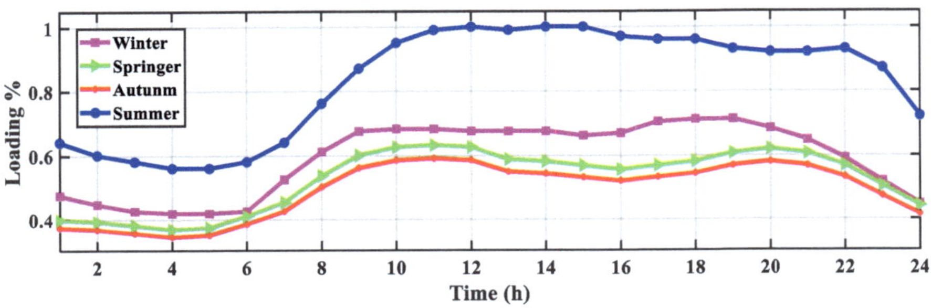


**Fig. 5.33** The hourly load profiles

Table 5.21, while the detailed data are given in [165]. The single line diagram of the network is depicted in Fig. 5.32 The system consists of 94 buses and 93 branches. The system load demand is 4797 kW+ j2323.900 kVAR. At 100% loading, the total active and reactive power losses are 365.173 kW and 505.785 kVar, respectively. Also, the minimum value of the voltage is 0.84749 p.u. at 92th bus. The captured results by MPA are compared with the PSO results. The empirical data of the MPA are adjusted to be $T_{max}$ equals 100 and the number of populations is 25. On the other hand, the data of the PSO is chosen to be $T_{max}$ equals 100, number of populations is 25, and $C1$ and $C2$ equal 2. The selected parameters of the GA are selected to be $T_{max}$ equals 100, the number of populations is 25, crossover percentage = 0.9, and mutation percentage = 0.2. The selected parameters of the SCA are selected to be $T_{max}$ equals 100, and the number of populations is 25. The program code of application the PSO, SCA, and GA for solving presented problem using m-file in MATLAB environmental. The cost coefficients are given in Table 5.7. Two case studies are presented where the hybrid PV-DG and D-STATCOM is incorporated under deterministic and probabilistic conditions as follows:

Two case studies are presented where the hybrid PV-DG and D-STATCOM is incorporated under deterministic and probabilistic conditions as follows: lines and buses data of this system are given in Appendix D.

## 5.6 Case 1: Allocation of the Hybrid System Without Considering the Uncertainty

It is well known that the load demand is varied hourly and seasonally due to variations of the customer activities and the solar irradiance is varied hourly and seasonally, which changes the output power of the PV units. The considered load variations are depicted in Fig. 5.33, which are given in [142], while the solar irradiance variations are shown in Fig. 5.34 [143]. The system power losses for four typical seasonal days at the base cased (without PV or D-STATCOM) are shown in Fig. 5.35. According to Fig. 5.35, the power losses follow or vary with load demand variations, where the highest power loss is being in the summer while the lowest power loss in the autumn. Two studied cases are presented including incorporating a single and two hybrid PV-DG and D-STATCOM systems as follows:

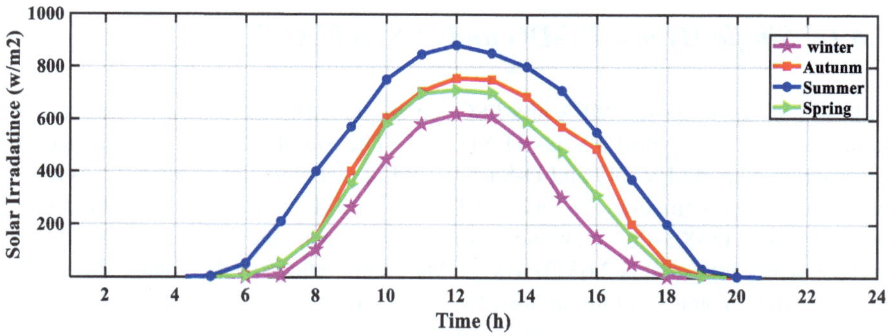


**Fig. 5.34** The hourly solar irradiance

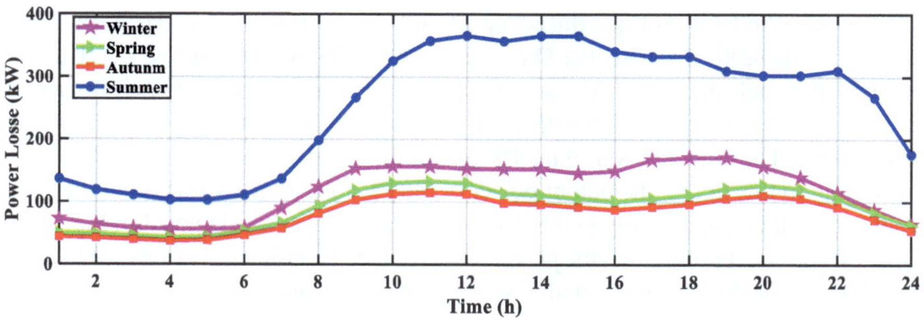


**Fig. 5.35** The system power losses

**Table 5.22** The simulation results of incorporating single hybrid PV-DG and D-STATCOM

| Item | Without hybrid system | MPA | PSO |
|---|---|---|---|
| **$E_{loss}$(kWh)** | 1.1888E+06 | 8.4498E+05 | 8.1143E+05 |
| **$E_{grid}$(kWh)** | 2.6728E+07 | 1.7273E+07 | 1.7615E+07 |
| Optimal location | – | 11 | 12 |
| **$P_{sr}$(kW)** | – | 4796 | 4598 |
| **$Q_{DS}$(kVAR)** | – | 2059 | 1908 |
| **TVD (p. u.)** | 4.6320E+04 | 1.2338E+04 | 1.2918E+04 |
| **TVSI (p. u.)** | 6.4856E+05 | 7.8938E+05 | 7.8403E+05 |
| **$C_{loss}$ ($)** | 7.1325E+04 | 5.0699E+04 | 4.8686E+04 |
| **$C_{Grid}$ ($)** | 2.5659E+06 | 1.6582E+06 | 1.6911E+06 |
| **$C_{PV}$ ($)** | – | 5.2489E+05 | 5.0322E+05 |
| **$C_{DS}$ ($)** | – | 1.0921E+04 | 1.0120E+04 |
| **$C_{Total}$ ($)** | 2.6372E+06 | 2.2447E+06 | 2.2531E+06 |

## *5.6.1 Single Hybrid PV-DG and D-STATCOM*

In this case, the proposed MPA and PSO are employed to assign the placement and rating of a single hybrid PV-D–STATCOM. The simulation results obtained by incorporating single PV-DG and D-STATCOM are tabulated in Table 5.22. Without including the combination PV-DG and D-STATCOM (base case), the total annual power loss is 1.1888E+06 kW, while the drawn power from the grid is 2.6728E+07 kW and the yearly total cost is 2.6372E+06 $. Also, summations of the voltage deviations (TVD) and the voltage stability index (TVSI) are 4.6320E+04 p.u. and 6.4856E+05 p. u., respectively. The best locations of the hybrid PV-DG and D-STATCOM that assigned by MPA and PSO are at the 11th bus and 12th bus, respectively. With optimal integration the hybrid PV-D-STATCOM, the total annual cost is reduced considerably to 2.2447E+06 $ and 2.2531E+06 $ using MPA and PSO techniques, respectively. In other words, the total cost is reduced by 14. 88% and 14.56% using MPA and PSO techniques, respectively. Also, the *TVD* is reduced to 1.2338E+04 p.u. and 1.2918E+04 p.u. with the application the MPA and PSO techniques, respectively. *TVD* is improved to 7.8938E+05 p.u. and 7.8403E+05 p.u. by employing the MPA and PSO techniques, respectively. The output power of the PV unit of the four seasons is shown in Fig. 5.36 According to this figure, the output power of the PV follows the variation of the solar irradiance. The typical hourly power losses for each season are depicted in Fig. 5.37. From this figure, it is clear that the power losses reduced considerably from 8:00 AM to 5:00 PM; this is due to the injected power by PV at this time.

## *5.6.2 Two Hybrid PV-DG and D-STATCOM Systems*

This section shows the results of the optimal inclusion of two hybrid PV-DG and D-STATCOM systems. The simulation results are tabulated in Table 5.23. Referring

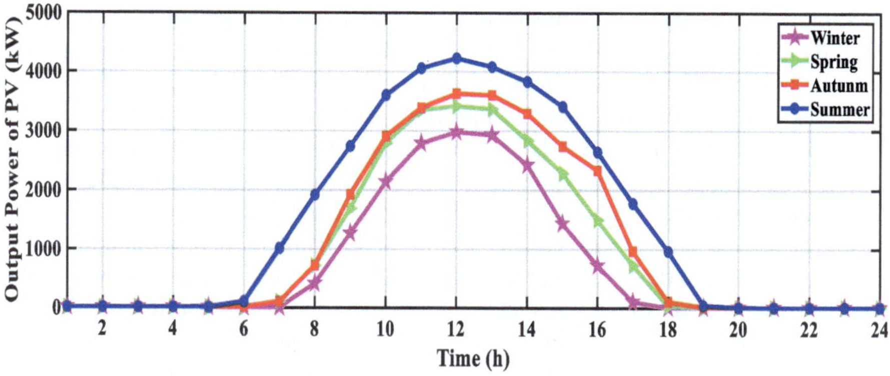


**Fig. 5.36** The hourly PV-DG output power

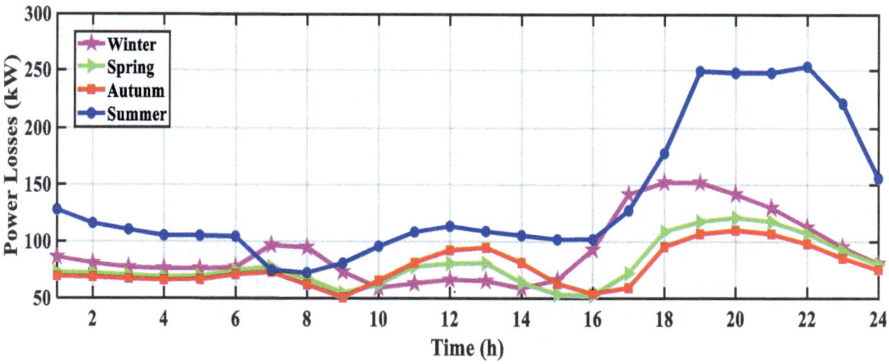


**Fig. 5.37** The system losses for case 1

**Table 5.23** The simulation results of incorporating two hybrid PV-DGs and D-STATCOMs

| Item | MPA | PSO |
|---|---|---|
| **$E_{loss}$(kWh)** | 7.3792E+05 | 8.7371E+05 |
| **$E_{grid}$(kWh)** | 1.7170E+07 | 1.8256E+07 |
| Optimal location1,<br>Optimal location 2 | 20,<br>53 | 47,<br>21 |
| **$P_{sr1}$(kW)**<br>**$P_{sr2}$(kW)** | 1063<br>3734 | 3123<br>1168 |
| **$Q_{DS1}$(kVAR)**<br>**$Q_{DS2}$(kVAR)** | 1133<br>1200 | 2210<br>563 |
| **TVD (p. u.)** | 1.0207E+04 | 1.0520E+04 |
| **TVSI (p. u.)** | 8.0432E+05 | 8.0462E+05 |
| **$C_{loss}$ (\$)** | 4.4275E+04 | 5.2422E+04 |
| **$C_{Grid}$ (\$)** | 1.6483E+06 | 1.7525E+06 |
| **$C_{PV}$ (\$)** | 5.2493E+05 | 4.6966E+05 |
| **$C_{DS}$ (\$)** | 6.0093E+03 | 1.1721E+04 |
| **$C_{Total}$ (\$)** | 2.2235E+06 | 2.2863E+06 |

to Table 5.23, the obtained results by employing MPA are better than those captured by PSO in terms of the total cost, voltage profile, and voltage stability. The optimal locations of the hybrid systems that have been assigned by MPA are at 20th bus and 53rd bus while the optimal sizes are listed in Table 5.23. Judging from Table 5.23, the inclusion of two hybrid PV-DG and D-STATCOM systems can reduce the total cost and enhance the system performance compared to the inclusion a single system. The total annual cost is reduced to 2.2235E+06 p.u. or by 15.69% compared with the base case as well as the TVD is reduced to 1.0207E+04 p.u. (77.96%), and the TVSI is increased to (24.02%) compared to the base case. The voltage profiles of the 94-bus system for seasons are depicted in Fig. 5.38. The voltage profile is enhanced with the hybrid system's inclusion, especially with the inclusion of two-hybrid PV-DG and D-STATCOM systems. Figures 5.39 and 5.40 illustrate, respectively, the injected powers variations of first PV-DG and second PV-DG. It can be noted that the powers of the PV-DGs track the solar irradiance variations in which the large variations start from 8:00 AM to 5:00 PM. Figure 5.40 describes the losses of the distribution system. It is evident that the drawn power from the grid and are system losses reduced considerably with the inserted powers by the PV systems.

Tables 5.24 and 5.25 summarize the obtained results and benefits of the inclusion of single and two hybrid systems of PV-DGs and D-STATCOM using different optimization algorithms. The obtained results by the application of MPA are superior compared with PSO, SCA, and GA in terms of the objective function. From Table 5.24, the total cost is reduced from 2.6372E+06 $ (without hybrid system) $ to 2.2447E+06 $ and 2.2235E+06 by incorporating single and two hybrid systems, respectively. Also, the total voltage deviation is reduced from 4.6320E+04 p.u. to 1.2338E+04 p.u. and 1.0207E+04 p.u., and the voltage stability is enhanced from 6.4856E+05 p.u. to 7.8938E+05 p.u. to 8.0432E+05 p.u.

## 5.7 Case 2: Allocation of the Hybrid System Considering the Uncertainty

In this case, the hybrid systems are incorporated considering the uncertainties of solar radiation and the connected load. The obtained load profile and the solar irradiance under uncertainties of the load demand and solar irradiance are illustrated in Figs. 5.41 and 5.42, respectively. The daily power losses are described in Fig. 5.43. The power losses vary with the load profile. The simulation results are listed in Table 5.26. Without incorporating PV-DG or D-STATCOM, the total annual cost, the summation of voltage deviation, and the stability index are 3.477094 E+06, $, 6.1228 E+04 p.u., and 6.00356 E+05 p.u., respectively. The studied cases are presented as follows:

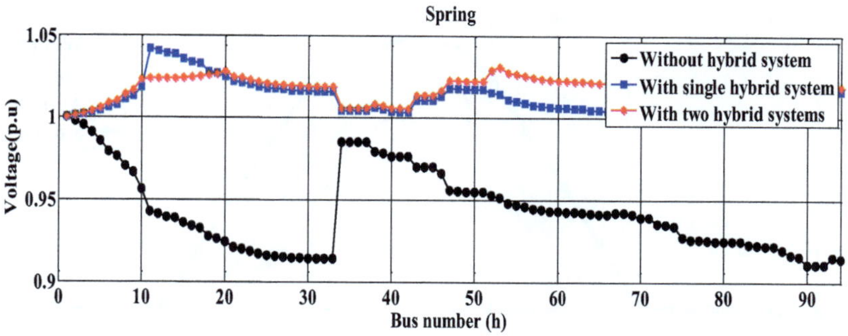


(a)

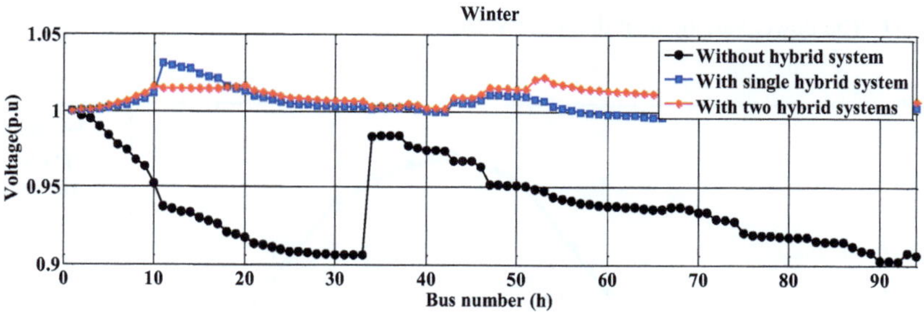


(b)

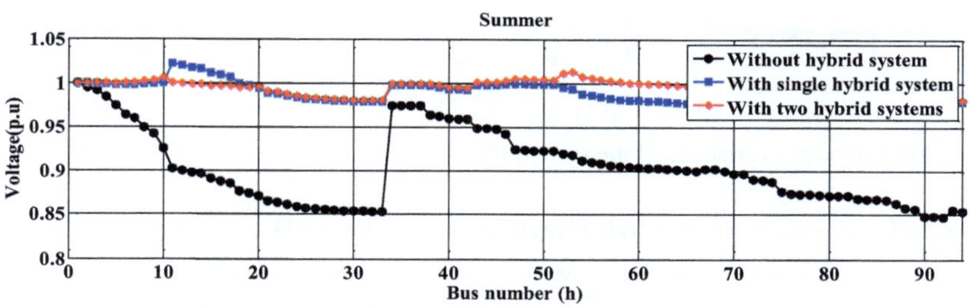


(c)

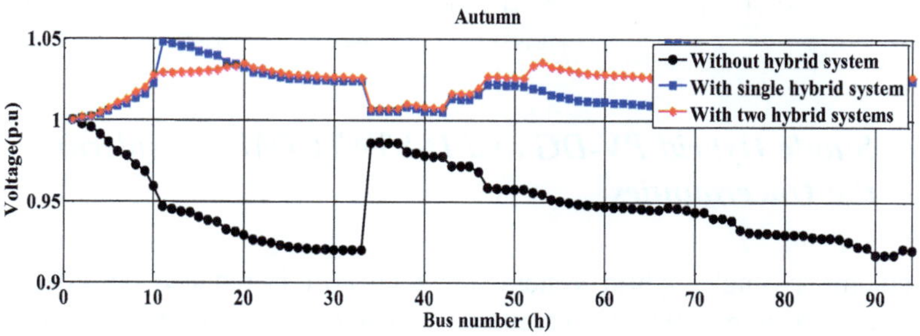


(d)

**Fig. 5.38** The voltage profile of system for (**a**) spring, (**b**) winter, (**c**) summer, and (**d**) autumn

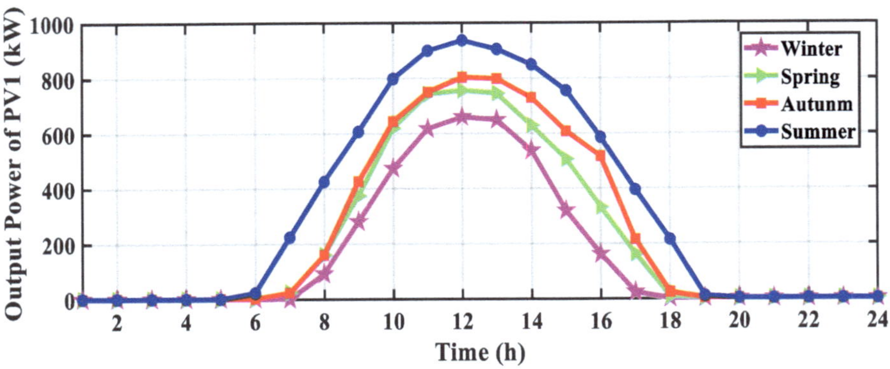


**Fig. 5.39** The power variations of the first PV-DG

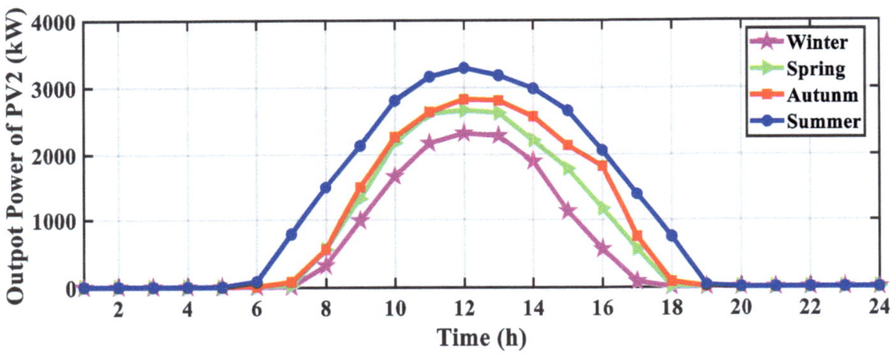


**Fig. 5.40** The power variations of the second PV-DG

**Table 5.24** The obtained results with a single hybrid PV-DG and D-STATCOM

| Item | Without hybrid system | MPA | PSO | SCA | GA |
|---|---|---|---|---|---|
| **TVD** (**p. u.**) | 4.6320E+04 | 1.2338E+04 | 1.2918E+04 | 1.5034E+04 | 1.3089E+04 |
| **TVSI**(**p. u.**) | 6.4856E+05 | 7.8938E+05 | 7.8403E+05 | 7.6902E+05 | 7.7477E+05 |
| $\mathbf{C_{Total}}$ ($) | 2.6372E+06 | 2.2447E+06 | 2.2531E+06 | 2.8794E+06 | 3.1224E+06 |
| **F** | | 0.4922 | 0.4969 | 0.6271 | 0.6626 |

## 5.7.1 Single Hybrid PV-DG and D-STATCOM Considering the Uncertainties

By including a single hybrid system, the total cost is reduced considerably to 3.0975E+06, $ or by 10.917% compared to without the insertion of PV-DG or D-STATCOM. Furthermore, it is reduced to 1.6816E+04 p.u. (72.53%), while it is increased to 7.6665E+05 p.u. (24.02%). Also, the drawn energy from the substation

**Table 5.25** The obtained results with two hybrid PV-DGs and D-STATCOMs

| Item | MPA | PSO | SCA | GA |
|---|---|---|---|---|
| **TVD (p. u)** | 1.0207E+04 | 1.0520E+04 | 1.6166E+04 | 1.3985E+04 |
| **TVSI (p. u)** | 8.0432E+05 | 8.0462E+05 | 7.5971E+05 | 7.7758E+05 |
| **$C_{Total}$ ($)** | 2.2235E+06 | 2.2863E+06 | 2.8734E+06 | 3.0275E+06 |
| **F** | 0.4766 | 0.4902 | 0.6320 | 0.6494 |

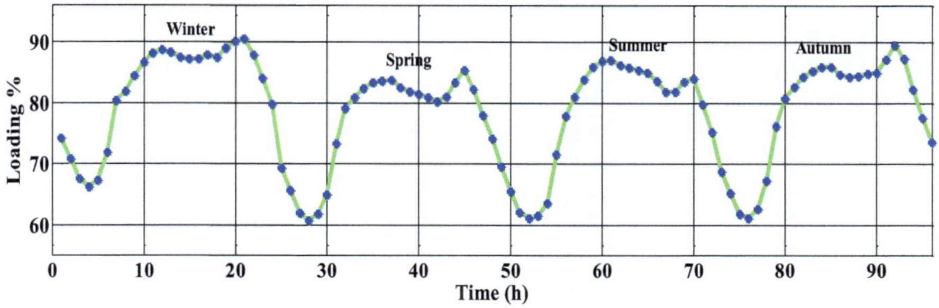


**Fig. 5.41** The load profile under uncertainty

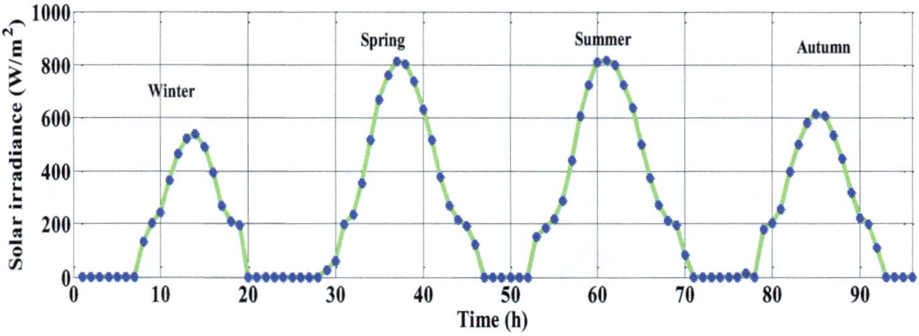


**Fig. 5.42** The solar irradiance under uncertainty

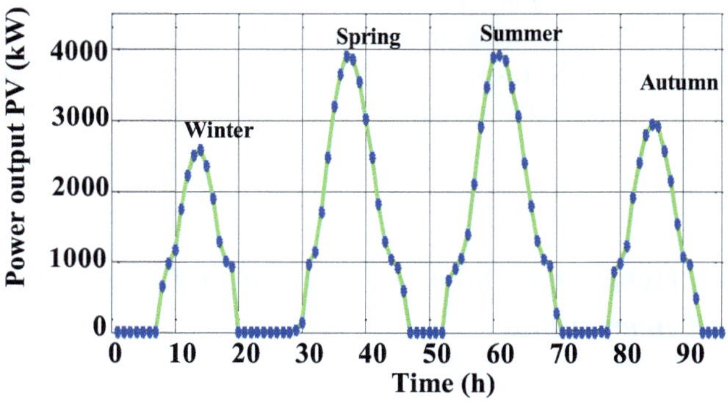


**Fig. 5.43** The output power of PV unit under the uncertainty condition

**Table 5.26** The simulation results of the inclusion of the hybrid PV-DG and D-STATCOMs under the uncertainties of system

| Item | Without hybrid system | Single hybrid system | Two hybrid systems |
|---|---|---|---|
| $E_{loss}$(kWh) | 1.905301 E+06 | 1.3088E+06 | 1.176065 E+06 |
| $E_{grid}$(kWh) | 3.5028922E +07 | 2.5912E+07 | 2.3540425 E+07 |
| Optimal location1,<br>Optimal location 2 | –<br>– | 15<br>– | 10<br>20 |
| $P_{sr1}$(kW)<br>$P_{sr2}$(kW) | –<br>– | 4797<br>– | 1629<br>3168 |
| $Q_{DS1}$(kVAR)<br>$Q_{DS2}$(kVAR) | –<br>– | 2323<br>– | 1565<br>1687 |
| **TVD (p. u.)** | 6.1228 E+04 | 1.6816E+04 | 8.179 E+03 |
| **TVSI (p. u.)** | 6.00356 E+05 | 7.6665E+05 | 8.10234 E+05 |
| $C_{loss}$ ($) | 1.14318 E+05 | 7.8529E+04 | 7.0563 E+04 |
| $C_{Grid}$ ($) | 3.362776 E+06 | 2.4876E+06 | 2.259880 E+06 |
| $C_{PV}$ ($) | – | 5.1906E+05 | 5.41452 E+05 |
| $C_{DS}$ ($) | – | 1.2321E+04 | 8.300 E+04 |
| $C_{Total}$ ($) | 3.477094 E+06 | 3.0975E+06 | 2.880197 E+06 |

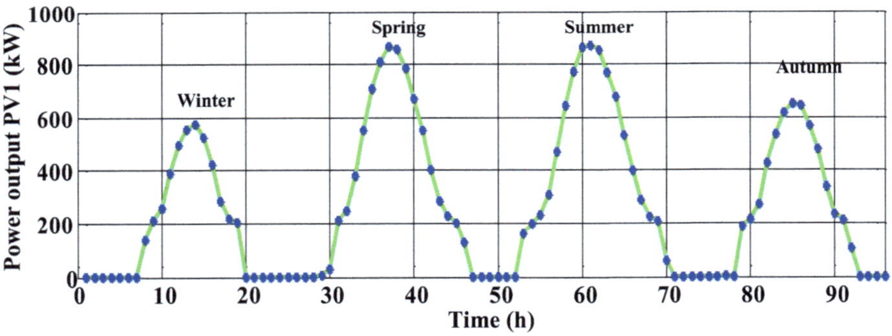


**Fig. 5.44** Variation of the first PV output power under uncertainty condition

and the energy losses are minimized considerably. The assigned best hybrid system position is on the 15th bus, while the ratings of the PV-DG and D-STATCOM are 4797 kW and 2323 kVAR, respectively. Variations of the PV power are depicted in Fig. 5.45. The variation in PV output power is caused by seasonal changes in solar irradiance.

## 5.7.2 *Two Hybrid PV-DG and D-STATCOM Considering the Uncertainties*

In this case, two hybrid systems are embedded in the system. According to Table 5.26, the total cost is reduced considerably to 2.880197E+06 $ or by

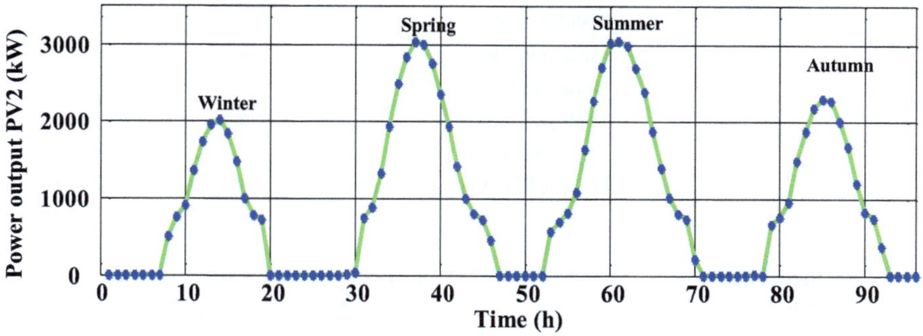


**Fig. 5.45** The output power of the second PV unit under uncertainty condition

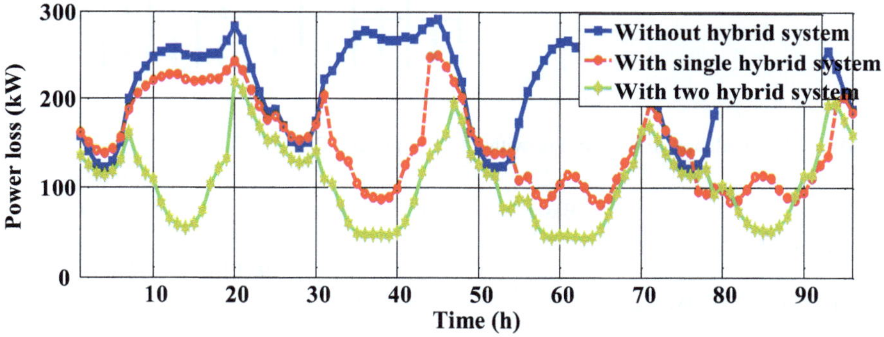


**Fig. 5.46** The power losses under uncertainty condition

17.16% compared to without insertion of PV-DG or D-STATCOM and is reduced to 8.179 E+03 p.u. (86.64%) and is increased to 8.10234E+05 p.u. (34.96%). The hybrid systems assigned the best locations are on the tenth and 20th buses, while the sizes of the first PV-DG and D-STATCOM are 3168 kW and 1565 kVAR, respectively. The second PV-DG and D-STATCOM have capacities of 1629 kW and 1687 kVAR, respectively. The first and second PV systems' output power are depicted in Figs. 5.44 and 5.45, respectively. Judging from Fig. 5.46, the power losses are reduced considerably with the inclusion of two hybrid PV-DGs and D-STATCOMs compared with a single hybrid system.

## 5.8 Optimal Allocation of DER in RDS Using SMA

### *5.8.1 IEEE-30 Bus Radial Distribution System*

In this section, numerical results are presented where the positioning and evaluation of PV modules and D-STATCOMs are made using SMA distribution in the East Delta Network (EDN) as part of the Unified Egyptian Network (UEN) network.

**Table 5.27** Parameters of system

| Specification | Value |
|---|---|
| Maximum iteration | 100 |
| Number of Population | 25 |
| Voltage limits | $0.90 \leq V_i \leq 1.05$ p. u. |
| PV sizing limits | $0 \leq P_{DG} \leq 23$ MW |
| D-STATCOM sizing limits | $0 \leq Q_{STATCOM} \leq 14$ MVar |

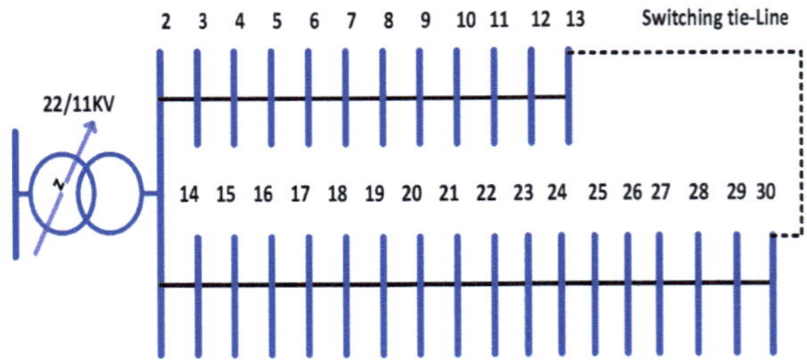


**Fig. 5.47** Single line diagram of the (EDN)

**Table 5.28** Simulation results with or without PV units and D-STATCOM

| Values | Basic | D-STATCOM | PV | PV & D-STATCOM |
|---|---|---|---|---|
| Optimal PV location (size kW) | – | – | 11 (484.5)<br>19 (14627) | 20 (11927)<br>7 (6676) |
| Optimal D-STATCOM location (size kVar) | – | 19 (13924)<br>6 (8327) | – | 16 (9878.9)<br>12 (1400) |
| Total $P_{PV}$ (kW) | – | – | 19,521,5 | 18,603 |
| Total $Q_{DSTATCOM}$ (KVar) | – | 22,251 | – | 11,278.9 |
| $V_{min}$ $(p.\ u.)$ @ bus no. | 0.94626 @ bus 30 | 0.96422 @ bus 30 | 0.99038 @ bus 30 | 0.99437 @ bus 30 |
| Total $P_{losses}$ (kW) | 805.733 | 790.5877 | 339.7244 | 123.0589 |
| Total $Q_{loss}$ (kVar) | 61.184 | 347.708 | 143.005 | 46.890 |
| VD (p. u.) | 1.0669 | 0.6698 | 0.1496 | 0.0652 |
| ∑VSI | 24.9867 | 26.4203 | 28.4080 | 28.8183 |

System constraints and SMA parameters are tabulated in Table 5.27. EDN system data are provided in [151]. Figure 5.47 shows the single EDN schematic diagram. The system load demand is 11 kV. The active system load and the reactive loads are 22,441,259 kW and 14,162,265 kV, respectively. The minimum voltage value is 0.94626 p.u. on bus 30. The total active loss is 805.733 kW. Three case studies were presented, including optimized assignment of D-STATCOMs only, optimized allocation of PV-only modules, and optimized assignment of D-STATCOM along with PV modules. All simulations are listed in Table 5.28. According to Table 5.30, the active power loss has decreased from 805.733 (base case) to 790.5877 kW, 339.7244 kW, and 123.0589 kW with optimal integration of D-STATCOMs only, and installation of PV modules only, and PV modules that were installed with

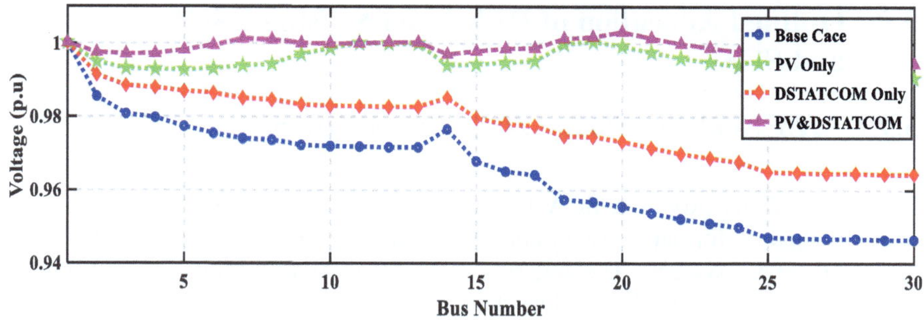


**Fig. 5.48** Voltage profile of all cases the system

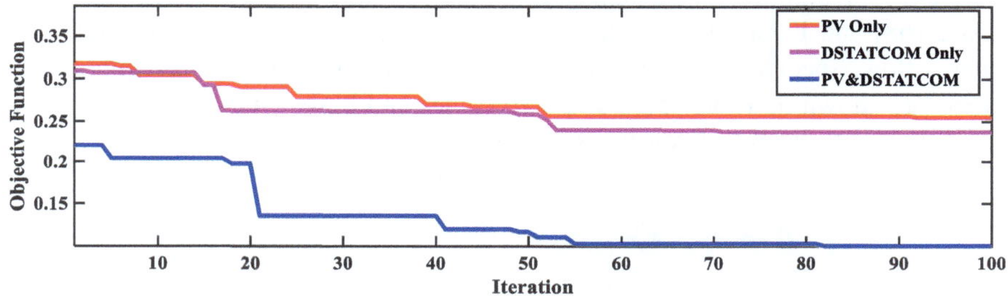


**Fig. 5.49** The convergence characteristic of the SMA for power losses minimization

D-STATCOMs, respectively. Lines and bus data for this system are given in Appendix B.

It can be seen from Table 5.28 that, voltage deviation is improved from 1.0669 p.u. (base case) to 0.6698 p.u. (when using D-STATCOM only), and is improved to 0.1496 p.u. (when using PV and D-STATCOM), and is improved to 0.0652 p.u. (when using PV and D-STATCOM). The voltage stability indicator has also been improved. Improved voltage profile from 24.9867 p.u. to 26.4203 (with D-STATCOM only), 28.4080 p.u. (with PV and D-STATCOM), and 28.8183 p.u. (with PV and D-STATCOM). From the results obtained, the system is greatly improved with the optimized allocation of PV and D-STATCOM simultaneously. This is compared with the optimized assignment of PV only or D-STATCOM. Figure 5.48 shows the voltage profile of the system, and it is evident that PV modulation with D-STATCOM can enhance the voltage profile greatly. Referring to Table 5.28, voltage stability also enhanced PV modulation with D-STATCOM. Figure 5.49 shows that the proposed algorithm has excellent and stable convergence characteristics as no fluctuation appears.

## 5.9 Optimal Allocation of DER in RDS Using ESCA and SCA

### *5.9.1 IEEE-33 Bus Radial Distribution System*

To determine the proposed SCA features and the enhanced SCA algorithm and examine their performance, two radial distribution systems were selected IEEE-33 and IEEE-69 bus systems. The load model for distribution systems is assumed to be a static power load [166]. The line diagram of a system considered is shown in Figs. 5.50 and 5.51. The proposed algorithm is a code using MATLAB 2019b and simulations are present on a Dell computer of an Intel core TM I7 processor with a frequency of 3.20 GHz and 32.0 GB of RAM. Back/forward sweep algorithm [167, 168] and its affinity are strong and assured [166] and solve calculations of energy flow. Specification's parameters and primary power flow results without any RESs are shown in Tables 5.28 and 5.29. Lines and buses data of this system are given in Appendix E and Appendix A (Table 5.30).

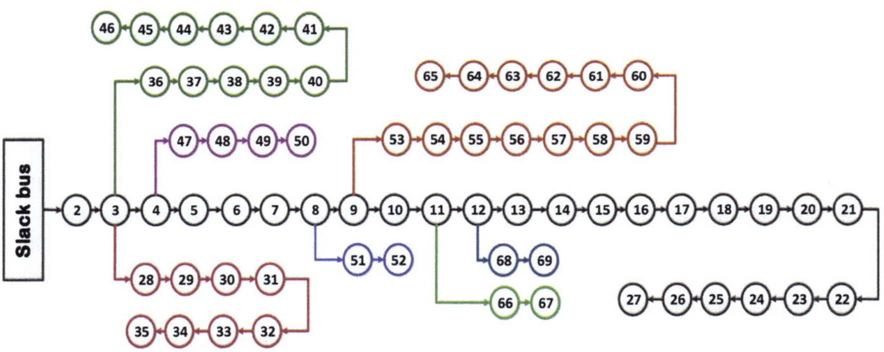


**Fig. 5.50** Single line diagram of IEEE-69 bus system

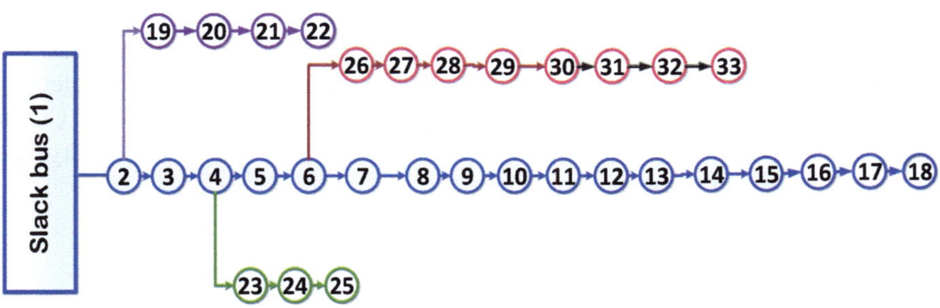


**Fig. 5.51** Single line diagram of IEEE-33 bus system

**Table 5.29** System specification and initial power flow

| Item | Value | Value |
|---|---|---|
| *System specifications:* | | |
| **NB** | 33 | 69 |
| **Npr** | 32 | 68 |
| Vsys (kV) | 12.66 | 12.66 |
| Base MVA | 100 | 100 |
| $\mathbf{S_{Load}}$**(MVA)** | 3.715 + j2.300 | 3.802 + j2.694 |
| $\mathbf{P_{Total_{loss}}}$ **(kW)** | 210.84 | 225 |
| $\mathbf{Q_{Total_{loss}}}$ **(kVar)** | 143.022 | 102.198 |
| $\mathbf{V_{min}}$ **(p. u.)** @ bus | 0.90378, 18 | 0.90919, 65 |

**Table 5.30** Input parameters used in numerical simulations

| Item | Set value(s) |
|---|---|
| *SCA-ESCA parameters:* | |
| Niter, max | 100 |
| Npop | 50 |
| Nruns | 50 |
| *System inequality constraints:* | |
| Bus voltage limits (p.u.) | ±5% |
| PV size limits (MVA) | $0 \leq SPV \leq 3$ |
| PV power factor limits [24] | $1 \leq PFPV \leq 1$ |

### 5.9.2 The First Case Study – 33-Bus System

The first test system of the proposed ESCA is the 33-bus system. Its total load is 3720 kW and 2300 kVA at a voltage level of 12.66 kV. Tables 5.31 and 5.32 describe the effects of the fixation of different types and numbers of PV on system performance. The superiority of the ESCA has been demonstrated in selecting the optimal sites and size of the photovoltaic compared to those obtained are shown in Tables 5.33, 5.34, and 5.35. Both PVs showed a significant reduction in active energy loss. In addition, a marked improvement in the voltage profile and system stability was achieved, as shown in Figs. 5.52 and 5.53, respectively. In the first method, bus number 6 has been suggested as the optimal location for PV. The optimum size for PV is 2590.2 kW. The power loss resulting from the proposed technology has decreased to 111.02 kW.

In the second method, the integration of multiple PV-DGs (two and three modules) is considered, and their effects are investigated. In the case of two PV-DGs, buses 13 and 30 were selected as ideal sites for PV-DG integration, with capacities of 851.4817 kW and 1154 kW, respectively. On the other hand, the optimal locations for three PV-DGs are buses 30, 13, and 24, and the PV capacities are 1053.5, 802, and 1091.3 kW, respectively. It is noteworthy that the active power loss was significantly reduced to 87.165 kW and 72.785 kW by two PV-DGs and three PV-DGs, respectively.

**Table 5.31** The results of installing of PV units in the first system

| Items | Without PV | With PV | | | | | |
|---|---|---|---|---|---|---|---|
| | | One PV | | Two PVs | | Three PVs | |
| | | ESCA | SCA | ESCA | SCA | ESCA | SCA |
| Total losses (kW) | 210.98 | 111.02 | 111.016 | 87.165 | 87.343 | 72.785 | 78.928 |
| Minimum voltage | 0.90378 @ bus 18 | 0.94237 @ bus 18 | 0.94231 @ bus 18 | 0.96839 @ bus 33 | 0.96626 @ bus 33 | 0.96867 @ bus 33 | 0.97711 @ bus 33 |
| Maximum voltage | 0.99703 @ bus 2 | 0.99858 @ bus 2 | 0.99858 @ bus 2 | 0.99826 @ bus 2 | 0.99822 @ bus 2 | 0.99882 @ bus 2 | 0.99910 @ bus 2 |
| PV size in kW (Location) | – | 2590.2 (6) | 2585.7 (6) | 851.4817 (13), 1154 (30) | 832.4591 (13), 1096 (30) | 1053.5 (30), 802 (13), 1091.3 (24) | 1100.8 (24), 1195 (12), 1151.0 (30) |
| VSI | 25.5401 | 28.5219 | 28.5166 | 29.3803 | 29.2442 | 29.6167 | 30.4819 |
| VD (p.u.) | 1.8044 | 0.9237 | 0.9252 | 0.6786 | 0.7156 | 0.6167 | 0.3865 |

### *5.9.3 The Second Case Study – 69-Bus System*

In this case, optimization results are obtained with ESCA mono- and multi-modulated PV, summarized in Table 5.36. The voltages and VSI voltage profiles are also illustrated in Figs. 5.54 and 5.55, respectively. The branch active energy loss profile is shown in Table 5.37. In the first method, renewable energy integration is considered as PV. Bus 61 was nominated as the best location for the installation of the 1872.7 kW PV array. The optimized layout of the PV module greatly reduces the active power loss and improves the voltage profile and the system stability so that the minimum bus voltage is 0.96829 p.u. and the VSI is 64.621 p.u.

In the second method, the optimal planning of renewable energy is examined considering the integration of multiple renewable energies. In the case of the two PV-DGs, bus numbers 18 and 61 were nominated as the best locations, and the optimum capacity of the PV-DGs is 531.26 kW and 1781 kW, respectively. It is noted that the input of two PV-DGs successfully participates in reducing the active power losses to 71.675 kW and increasing the minimum voltage point to 0.97893 p.u. and VSI to 66.021 p.u. where the active power loss is reduced to 69.449 kW with three PV-DG power stations. Moreover, the system bus voltage profile was significantly improved by a precise uniform pattern, as in Fig. 5.55, which was confirmed by an improvement in the minimum value of bus voltages before and after connection. The best value of power losses (for a moment, 83,224 kW with one PV) obtained by the proposed algorithm is lower than the best values shown in previous work. This reveals the efficiency of the algorithm developed over other methods to solve the problem of optimal allocation of renewable energy in distribution networks. Tables 5.38, 5.39, and 5.40 are examples of this. Comparative results of

**Table 5.32** Results of optimized allocation of PV units in 33 bus

| Without PV | | With multi-objective function | | | | | |
|---|---|---|---|---|---|---|---|
| | | ESCA | | | SCA | | |
| Items | | PV1 | PV2 | PV3 | PV1 | PV2 | PV3 |
| Total losses (kW) | 210.98 | 111.012 | 87.252 | 72.885 | 111.016 | 87.560 | 76.664 |
| Minimum voltage | 0.90378 @ bus 18 | 0.94237 @ bus 18 | 0.96425 @ bus 18 | 0.96772 @ bus 33 | 0.94231 @ bus 18 | 0.96533 @ bus 18 | 0.95496 @ bus 18 |
| Maximum voltage | 0.99703 @ bus 2 | 0.99858 @ bus 2 | 0.99830 @ bus 2 | 0.99882 @ bus 2 | 0.99858 @ bus 2 | 0.99835 @ bus 2 | 0.99857 @ bus 2 |
| PV size in kW (Location) | – | 2590.2 (6) | 1104.4(30), 957 (12) | 11,091 (24), 838 (13), 1006.2(30) | 2585.7 (6) | 947.3237 (12), 1215 (30) | 795.8191(11), 831 (31), 896.2417 (24) |
| VSI (p.u.) | 25.5401 | 28.5219 | 29.7753 | 29.6395 | 28.5166 | 29.6288 | 29.0059 |
| VD (p.u.) | 1.8044 | 0.9237 | 0.5720 | 0.6106 | 0.9252 | 0.6126 | 0.7840 |

**Table 5.33** Comparative results for incorporating single PV in 33-bus system

| Technique | Without | GA [169] | PSOPC [170] | EVPSO [170] | AEPSO [170] | ADPSO [170] | DAPSO [170] | Analytical [153] | GA [160] |
|---|---|---|---|---|---|---|---|---|---|
| $P_{loss}$(KW) | 210.98 | 105.481 | 136.75 | 140.19 | 131.43 | 129.53 | 127.17 | 111.24 | 132.64 |
| Location | – | 6 | 15 | 11 | 14 | 13 | 8 | 6 | 6 |
| Size (kVA) | – | 2580 | 763 | 1000 | 1200 | 1210 | 1212 | 2490 | 2380 |
| Technique | BSOA [28] | ESCA | | | | | | | |
| $P_{loss}$(KW) | 118.12 | 111.02 | | | | | | | |
| Location | 8 | 6 | | | | | | | |
| Size (kVA) | 1857.5 | 2590.2 | | | | | | | |

**Table 5.34** Comparative results for incorporating two PVs in 33-bus system

| Technique | Without | MINLP [159] | Exhaust-ve OPF [154] | EAOPF [154] | EA [154] | Hybrid [157] | PSO [157] | IA [171] | ESCA |
|---|---|---|---|---|---|---|---|---|---|
| $P_{loss}$(KW) | 210.98 | 87.167 | 87.17 | 87.17 | 87.172 | 87.280 | 87.170 | 87.550 | 87.165 |
| Location | – | 13<br>30 | 13<br>30 | 13<br>30 | 13<br>30 | 13<br>30 | 13<br>30 | 12<br>30 | 13<br>30 |
| Size (kVA) | – | 850,<br>1150 | 852,<br>1158 | 852,<br>1158 | 844,<br>1149 | 830,<br>1110 | 850,<br>1160 | 1020,<br>1020, | 851.4817,<br>1154 |

**Table 5.35** Comparative results for incorporating three PVs in 33-bus system

| Technique | Without | MINLP [159] | Exhaustive OPF [154] | EA-OPF [154] | EA [154] | Hybrid [157] | PSO [157] | IA [171] | ESCA |
|---|---|---|---|---|---|---|---|---|---|
| $P_{loss}$(KW) | 210.98 | 105.481 | 136.75 | 140.19 | 131.43 | 129.53 | 127.17 | 111.24 | 132.64 |
| Location | – | 13<br>24<br>30 | 13<br>24<br>30 | 13<br>24<br>30 | 13<br>24<br>30 | 13<br>24<br>30 | 13<br>24<br>30 | 13<br>24<br>30 | 30<br>13<br>24 |
| Size (kVA) | – | 800,<br>1090,<br>1050 | 802,<br>1091,<br>1054 | 802,<br>1091,<br>1054 | 798,<br>1099,<br>1050 | 790,<br>1070,<br>1010 | 770,<br>1090,<br>1070 | 900,<br>900,<br>900 | 1053.5,<br>802,<br>1091.3 |

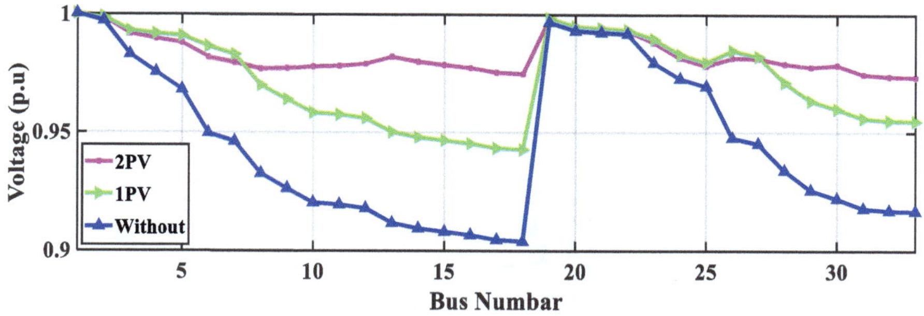


**Fig. 5.52** Voltage profile of the 33-bus system

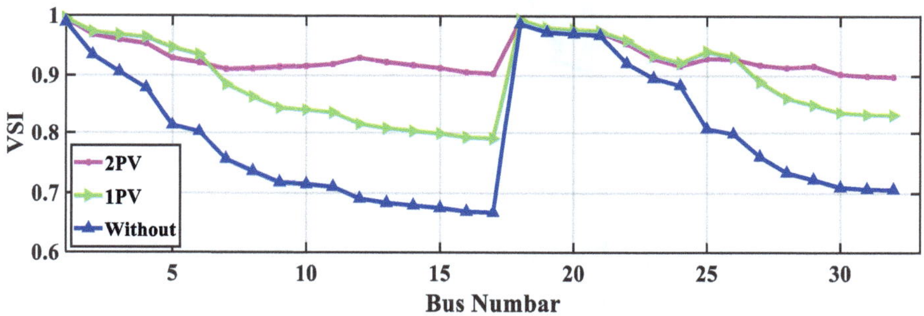


**Fig. 5.53** Voltage stability index of the 33-bus system

**Table 5.36** The results of installing of PV units in the second system

| Items | Without PV | With PV | | | | | |
|---|---|---|---|---|---|---|---|
| | | One PV | | Two PVs | | Three PVs | |
| | | ESCA | SCA | ESCA | SCA | ESCA | SCA |
| Total losses (kW) | 225 | 83.222 | 83.231 | 71.675 | 72.201 | 69.484 | 70.757 |
| Minimum voltage | 0.90919 @ bus 65 | 0.96829 @ bus 27 | 0.96819 @ bus 27 | 0.97893 @ bus 65 | 0.98169 @ bus 65 | 0.9793 @ bus 65 | 0.974 @ bus 65 |
| Maximum voltage | 0.99997 @ bus 2 | 0.99997 @ bus 2 | 0.99997 @ bus 2 | 0.99997 @ bus 2 | 0.9999 @ bus 2 | 0.9999 @ bus 2 | 0.999 @ bus 2 |
| PV size in kW (Location) | – | 1872.7 (61) | 1856 (61) | 531.259 (18), 1781 (61) | 1856 (61), 566 (16) | 596 (11), 381 (18), 1719 (61) | 1601 (61), 420 (19), 465.01 (9) |
| VSI (p.u.) | 61.2181 | 64.6212 | 64.5925 | 66.0295 | 66.3084 | 66.3258 | 65.8222 |
| VD (p.u.) | 1.8374 | 0.8729 | 0.8805 | 0.5002 | 0.4280 | 0.4241 | 0.5549 |

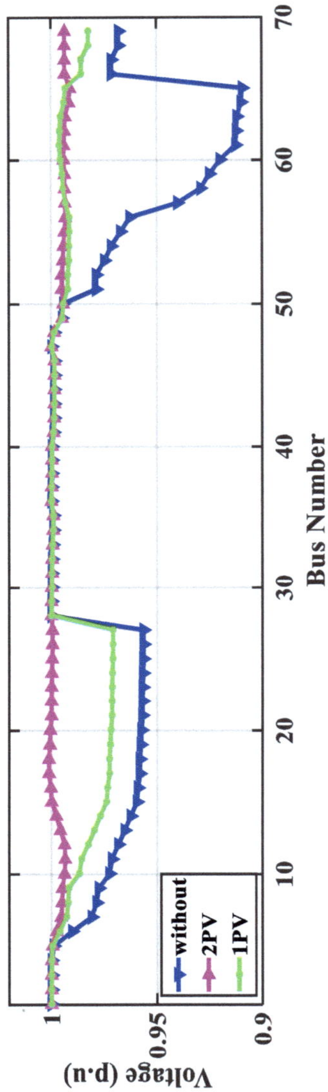


**Fig. 5.54** Voltage profile of the 69-bus system

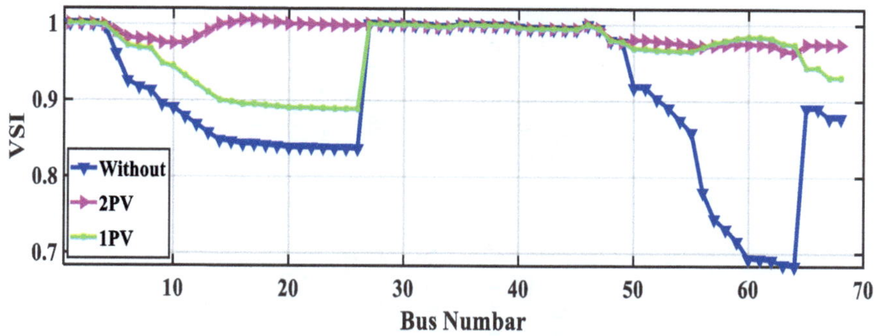


**Fig. 5.55** Voltage stability index of the 69-bus system

incorporating PV into a 69-bus system are shown in Figs. 5.56, 5.57, 5.58, 5.59, 5.60, and 5.61. From the affinity graph, the objective value (total energy loss) converges rapidly in ESCA.

### *5.9.4 Optimal Integral of PV-DG along with D-STATCOMS*

In this case, PV-DGs and D-STATCOMs are inserted into the IEEE-33 and IEEE-69 bus systems. The results gained for these cases are listed in Table 5.41. The optimal positions and classifications of the PV-DGs and D-STATCOMs are also qualified. Referring to Table 5.41, it can be evident that the active power loss has been significantly reduced with 33 bus systems, including one PV with one D-STATCOM to 47.059 kW and two PVs with two D-STATCOMs to 20.881 kW, respectively. Judging from Table 5.42, it is evident that the voltage profile and system stability were greatly improved with the simultaneous inclusion of D-STATCOMs and PV-DGs compared to PV inclusion only. Figures 5.62 and 5.63 show the voltage profile of the system with the input of D-STATCOMs and PV-DGs. Figures 5.64 and 5.65 illustrate the VSI voltage deviations of the system with the introduction of PV D-STATCOMs. Referring to Table 5.42, it can be evident that the active power loss has decreased significantly with 69-bus systems, comprising one PV with one D-STATCOM to 14.360 kW and two PVs with two D-STATCOMs to 9.265 kW, respectively.

**Table 5.37** Results of optimal allocation of PV-DGs

| Without PV | | With multi-objective function | | | | | |
|---|---|---|---|---|---|---|---|
| | | ESCA | | | SCA | | |
| Items | | PV1 | PV2 | PV3 | PV1 | PV2 | PV3 |
| Total losses (kW) | 225 | 89.230 | 77.605 | 74.412 | 89.393 | 79.677 | 77.635 |
| Minimum voltage | 0.90919 @ bus 65 | 0.97072 @ bus 27 | 0.99087 @ bus 65 | 0.98810 @ bus 65 | 0.97075 @ bus 27 | 0.99211 @ bus 65 | 0.99179 @ bus 65 |
| Maximum voltage | 0.99997 @ bus 2 | 0.99997 @ bus 2 | 1.00118 @ bus 17 | 0.99998 @ bus 2 | 0.99997 @ bus 2 | 1.00298 @ bus 15 | 0.99998 @ bus 2 |
| PV size in kW (Location) | – | 2293.4 (61) | 743.7399 (18), 2096 (61) | 576.8192 (54), 1900 (61), 601.1664 (22) | 2299.1 (61) | 827.4227 (16), 2119 (61) | 2128.2 (61), 716 (16), 705.0186 (3) |
| VSI (p.u.) | 61.2181 | 65.3700 | 67.2296 | 67.1408 | 65.3645 | 66.9367 | 67.1311 |
| VD (p.u.) | 1.8374 | 0.6766 | 0.2002 | 0.2150 | 0.6780 | 0.2672 | 0.2170 |

**Table 5.38** Comparative results for incorporating single PV in 69-bus system

| Technique | Without | MINLP [159] | Exhaustive OPF [154] | EA-OPF [154] | HACO [172] | Hybrid [157] | PSO [157] | GA [158] | GA [160] |
|---|---|---|---|---|---|---|---|---|---|
| $P_{loss}$(KW) | 225 | 83.222 | 83.23 | 83.23 | 83.323 | 83.222 | 83.372 | 83.222 | 132.64 |
| Location | – | 61 | 61 | 61 | 61 | 61 | 61 | 61 | 61 |
| Size (kVA) | – | 1870 | 1870 | 1870 | 1872 | 1810 | 1870 | 1794 | 1819.691 |
| Technique | ESCA | | | | | | | | |
| $P_{loss}$(KW) | 83.222 | | | | | | | | |
| Location | 61 | | | | | | | | |
| Size (kVA) | 1872.7 | | | | | | | | |

**Table 5.39** Comparative results for incorporating two PVs in 69-bus system

| Technique | Without | MINLP [159] | GA [160] | CSA [156] | SGA [156] | PSO [156] | MTLBO [155] | GA [158] | ESCA |
|---|---|---|---|---|---|---|---|---|---|
| $P_{loss}$(KW) | 225 | 71.693 | 71.7912 | 76.4 | 82.9 | 78.8 | 71.776 | 84.233 | 71.675 |
| Location | – | 17<br>61 | 61<br>11 | 22<br>61 | 17<br>61 | 14<br>62 | 17<br>61 | 1<br>62 | 18<br>61 |
| Size (kVA) | – | 510,<br>1780 | 1777,<br>555 | 600,<br>2100 | 1000,<br>2400 | 700,<br>2100 | 519.705,<br>1732.004 | 6,<br>1794 | 531. 2598,<br>1781 |

**Table 5.40** Comparative results for incorporating three PVs in 69-bus system

| Technique | Without | EA [154] | MTLBO [155] | KHA [161] | Hybrid [157] | PSO [156] | ESCA |
|---|---|---|---|---|---|---|---|
| $P_{loss}$(KW) | 225 | 69.62 | 69.539 | 69.56 | 69.52 | 69.541 | 69.484 |
| Location | – | 11<br>18<br>61 | 11<br>18<br>61 | 12<br>22<br>61 | 11<br>17<br>62 | 11<br>17<br>61 | 11<br>18<br>61 |
| Size (kVA) | – | 467,<br>380,<br>1795 | 493,<br>378,<br>1672 | 496,<br>311,<br>1735 | 510,<br>380,<br>1670 | 460,<br>440,<br>1700 | 596.0304,<br>381,<br>1719.0 |

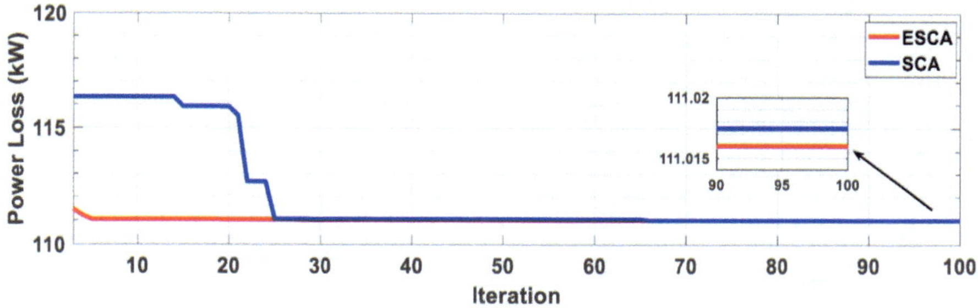


**Fig. 5.56** Change of total power loss single PV with iterations for the 33-bus system

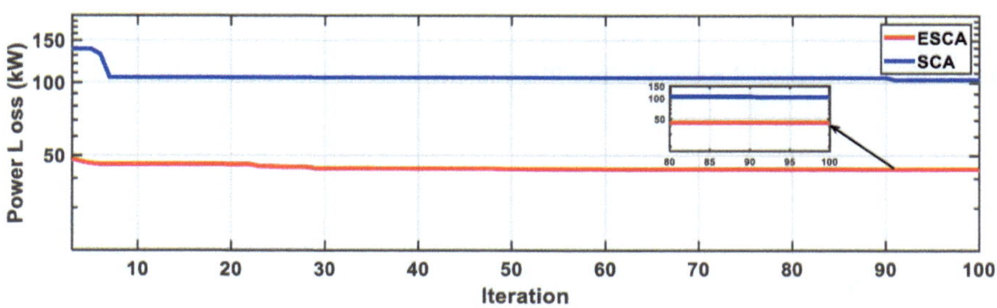


**Fig. 5.57** Change of total power loss two PVs with iterations for the 33-bus system

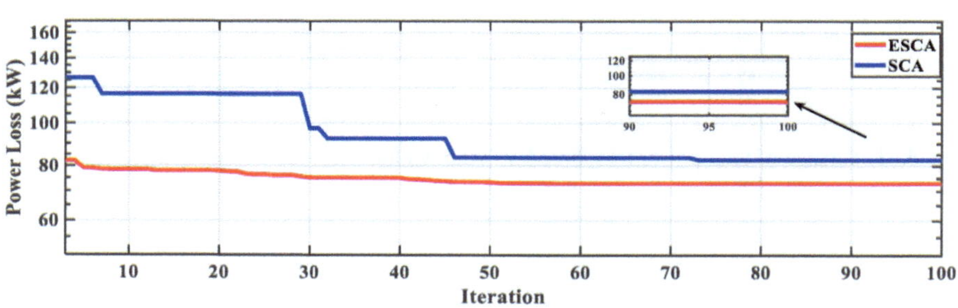


**Fig. 5.58** Change of total power loss three PVs with iterations for the 33-bus system

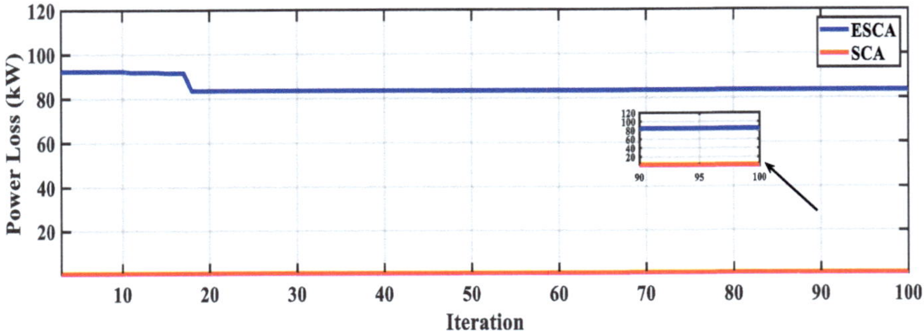


**Fig. 5.59** Change of total of total power loss single PV with iterations for the 69-bus system

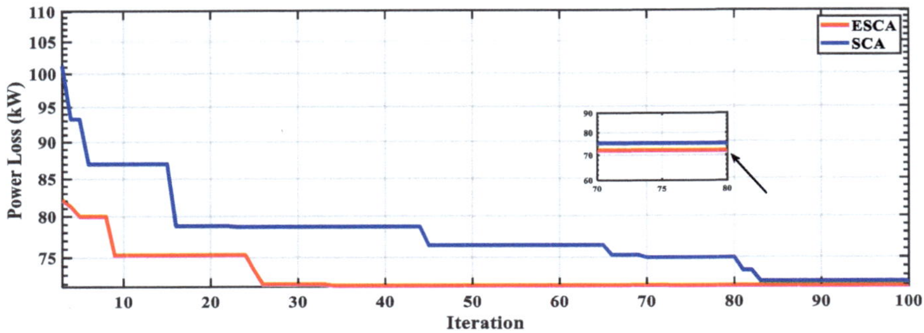


**Fig. 5.60** Change of total power loss two PVs with iterations for the 69-bus system

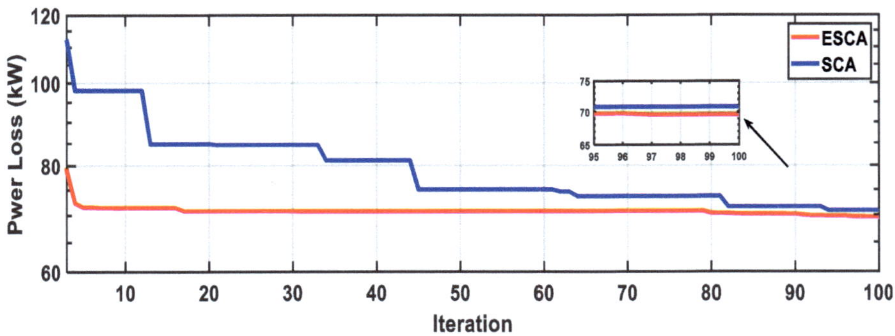


**Fig. 5.61** Change of total power loss three PVs with iterations for the 69-bus system

**Table 5.41** Simulation result with inclusion PV and D-STATCOM

| Items | Without –PV 33 | With multi-objective function | | | |
|---|---|---|---|---|---|
| | | ESCA | | SCA | |
| | | Single PV & single D-STATCOM | Two PVs & two D-STATCOMs | Single PV & single D-STATCOM | Two PVs & two D-STATCOMs |
| Total losses (kW) | 210.98 | 58.443 | 28.882 | 64.024 | 30.265 |
| Total reactive loss (KVAR) | | 47.059 | 20.881 | 49.646 | 21.736 |
| Minimum voltage | 0.90378 @ bus 18 | 0.95365 @ bus 18 | 0.98075 @ bus 25 | 0.94658 @ bus 18 | 0.97995 @ bus 18 |
| Maximum voltage | 0.99703 @ bus 2 | 0.99896 @ bus 2 | 1.00382 @ bus 30 | 0.99869 @ bus 2 | 0.99918 @ bus 30 |
| PV size in kW (Location) | – | 2531.7 (6) | 873.7099 (13), 1192 (30) | 2188.7 (6) | 743 (12), 1167 (30) |
| D-STATCOM size in KVA (location) | – | 1256 (30) | 460 (23), 1077 (14) | 1015 (29) | 460 (5), 980 (2) |
| VSI (p.u.) | 25.5401 | 29.8044 | 31.5515 | 29.1186 | 30.8336 |
| VD (p.u.) | 1.8044 | 0.5738 | 0.1453 | 0.7590 | 0.2969 |

**Table 5.42** Simulation result with inclusion PV and D-STATCOM

| Items | Without –PV 69 | With multi-objective function | | | |
|---|---|---|---|---|---|
| | | ESCA | | SCA | |
| | | Single PV & single D-STATCOM | Two PVs & two D-STATCOMs | Single PV & single D-STATCOM | Two PVs & two D-STATCOMs |
| Total losses (kW) | 225 | 23.441 | 11.773 | 26.734 | 12.477 |
| Total reactive loss (KVAR) | | 14.360 | 9.265 | 16.434 | 10.084 |
| Minimum voltage | 0.90919 @ bus 65 | 0.97306 @ bus 27 | 0.99427 @ bus 50 | 0.97054 @ bus 27 | 0.99313 @ bus 69 |
| Maximum voltage | 0.99997 @ bus 2 | 1.00260 @ bus 62 | 1.00325 @ bus 14 | 0.99998 @ bus 2 | 0.99999 @ bus 2 |
| PV size in kW (location) | – | – 1902.2 (61) | 1757.5 (61), 835 (13) | 1.6430 (61) | 380 (21), 1636 (61) |
| D-STATCOM size in KVA (location) | – | 1353 (61) | 1052.8 (10), 757 (56) | 1037 (61) | 538 (36), 1430 (3) |
| VSI (p.u.) | 61.2181 | 65.8998 | 67.8398 | 26.7343 | 67.3390 |
| VD (p.u.) | 1.8374 | 0.5617 | 0.0755 | 0.7332 | 0.1661 |

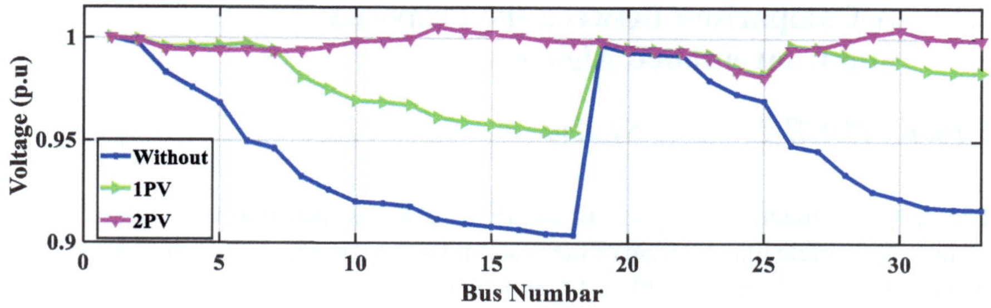


**Fig. 5.62** Voltage profile with 33-bus system

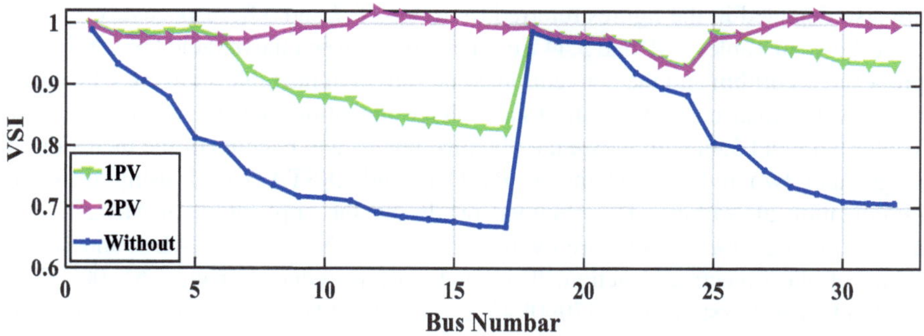


**Fig. 5.63** Voltage stability index with 33-bus system

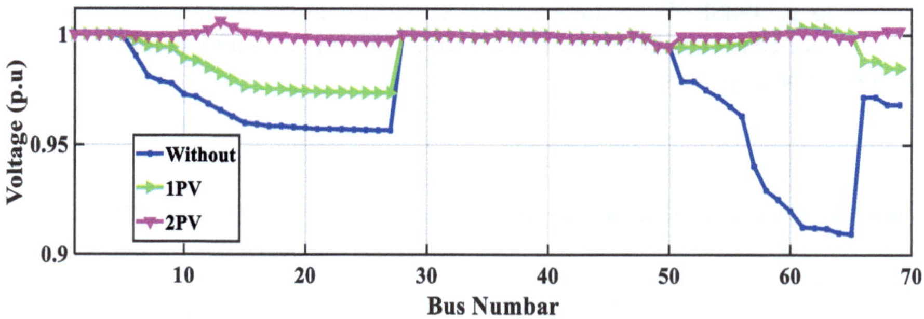


**Fig. 5.64** Voltage profile 69-bus system

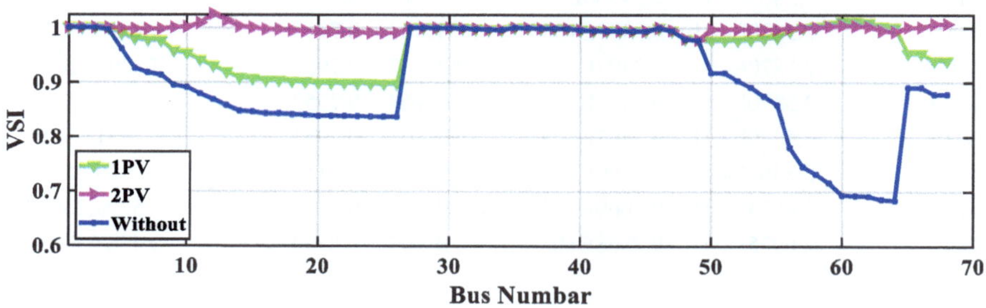


**Fig. 5.65** Voltage stability index with 69-bus system

## 5.10 A Comparison Between the Proposed and Conventional Algorithms

### *5.10.1 IEEE-118 Bus System*

The proposed method is applied to the IEEE 118-bus distribution as a large-scale system. The single line diagram of the system is shown in Fig. 5.20, while the system specifications are listed in Table 5.11, and the line and bus data are given in [173]. In this case, the proposed algorithm is applied to solve the optimal power planning problem under uncertainty conditions for the 118-bus system as a large-scale system. The system load profile and solar irradiance for this case under uncertain conditions are also depicted in the figures referred to in Figs. 5.7 and 5.8, respectively.

It shows in Tables 5.40 and 5.44 a comparison between the obtained results by the proposed algorithm and the conventional one shown in the table showing which one is better for total cost. Also, the total voltage deviation and the voltage stability Tables 5.43 and 5.44 summarize the obtained results and benefits of the inclusion of single and two hybrid systems of PV-DGs and D-STATCOM using different optimization algorithms. The obtained results by the application of MALO, EO, and LAPO are superior compared with WOA, SMA, MPA, SCA, ALO, and GA in terms of the objective function. The total cost is reduced from 1.61300E+7 $ (without hybrid system) to 1.47990E+7 $ and 1.465397E+7 $ with single and two hybrid systems using MALO, respectively, according to Tables 5.43 and 5.44. Also, the total voltage deviation is reduced from 3.5605E+4 p.u. to 2.5869E+4 p.u. and 2.1244E+4 p.u. using MALO, and the voltage stability is enhanced from 8.93320E +5 p.u. to 9.34980E+5 p.u. to 9.51969E+5 p.u. Figures 5.66 and 5.67 show the convergence characteristics of the considered algorithms with integration of a single and two hybrid systems.

**Table 5.43** Comparative results for incorporating single hybrid PV and D-STATCOM in 118-bus system

| Single hybrid system | TVD (p. u.) | TVSI (p. u.) | $C_{\text{Total}}$ ($) | Objective function | Elapsed time is seconds |
|---|---|---|---|---|---|
| Base case | 3.5605E+4 | 8.93320E+5 | 1.61300E+7 | 1 | 9.481155 |
| ALO | 2.5614E+4 | 9.33530E+5 | 1.58790E+7 | 0.8387 | 969.608573 |
| MALO | 2.5869E+4 | 9.34980E+5 | 1.47990E+7 | 0.7957 | 364.597171 |
| EO | 2.5870E+04 | 9.3497E+05 | 1.4799E+07 | 0.7957 | 4926.863778 |
| LAPO | 2.5870E+04 | 9.3497E+05 | 1.4799E+07 | 0.7957 | 8391.381205 |
| MPA | 2.5870E+04 | 9.3497E+05 | 1.4799E+07 | 0.7957 | 6562.739330 |
| SMA | 2.5870E+04 | 9.3497E+05 | 1.4799E+07 | 0.7957 | 2851.988515 |
| WOA | 2.8367E+04 | 9.4069E+05 | 1.4895E+07 | 0.8154 | 2929.457210 |
| SCA | 2.8574E+04 | 9.4069E+05 | 1.4895E+07 | 0.7976 | 5299.289890 |
| GA | 2.6975E+04 | 9.2868E+05 | 1.5664E+07 | 0.8393 | 478.630037 |

**Table 5.44** Comparative results for incorporating two hybrid PV and D-STATCOM in 118-bus system

| Two hybrid system | TVD (p. u.) | TVSI (p. u.) | $C_{\text{Total}}$ (\$) | Objective function | Elapsed time is seconds |
|---|---|---|---|---|---|
| Base case | 3.5605E+4 | 8.93320E+5 | 1.61300E+7 | 1 | 9.481155 |
| ALO | 2.4357E+4 | 9.40625E+5 | 1.53496E+7 | 0.8079 | 3478.563853 |
| MALO | 2.1244E+4 | 9.51969E+5 | 1.465397E+7 | 0.75719 | 3478.563853 |
| EO | 2.56813E+04 | 9.30537E+05 | 1.46378E+07 | 0.7876 | 5555.499277 |
| LAPO | 2.0812E+04 | 9.51717E+05 | 1.47960E+07 | 0.76005 | 8469.623613 |
| MPA | 1.9023E+04 | 9.62739E+05 | 1.48466E+07 | 0.74959 | 6614.239384 |
| SMA | 1.9242E+04 | 9.60344E+05 | 1.472207E+07 | 0.7459 | 2851.988515 |
| WOA | 2.0776E+04 | 9.51111E+05 | 1.60717E+07 | 0.81273 | 3034.870266 |
| SCA | 2.4712E+04 | 9.37086E+05 | 1.64404E+07 | 0.85566 | 5335.752573 |
| GA | 1.9888E+04 | 9.56592E+05 | 1.55868E+07 | 0.78637 | 3766.212795 |

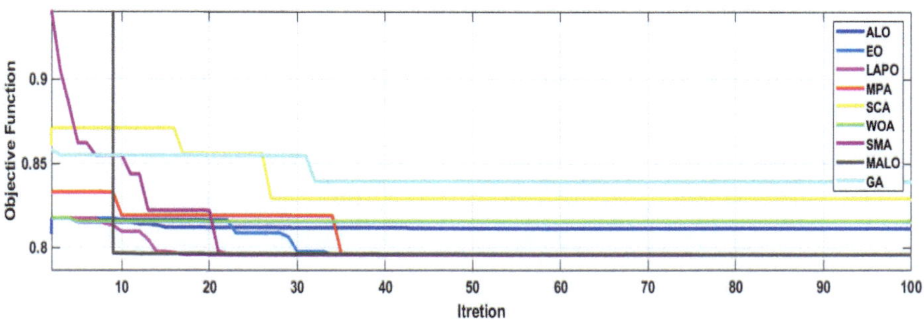


**Fig. 5.66** The convergence characteristics of the considered algorithms with integration single hybrid systems

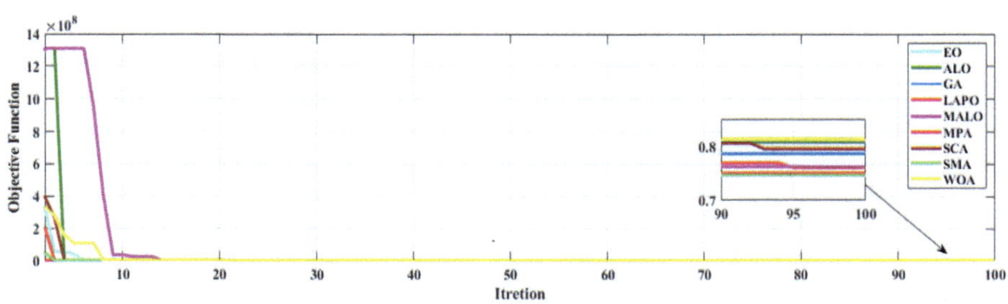


**Fig. 5.67** The convergence characteristics of the considered algorithms with integration two hybrid systems

# Chapter 6
# Conclusions and Future Work

## 6.1 Conclusions

Several benefits can be captured by optimizing the photovoltage distributed generation (PV-DG) and the distributed static compensator (D-STATCOM), where the PV-DG can provide active power to the load. In contrast, D-STATCOM can provide reactive power to the connected load. However, the inclusion of the PV-DG and the D-STATCOM is a strenuous task due to the continuous variations in the load and the solar irradiance.

*This book* deals with the integration of the PV-DG and the D-STATCOM in radial distribution networks (RDN) to improve the system performance and reduce costs. The optimal sites and sizes of the PV-DGs and the D-STATCOMs have been determined using developed optimizers such as marine predators algorithm (MPA), equilibrium optimizer (EO), lightning attachment procedure optimization (LAPO), sine cosine algorithm (SCA), ant lion optimizer (ALO), whale optimization algorithm (WOA), and slime mold algorithm (SMA). In this work, the forward backward method is used for solving the power flow in RDN, and the candidate buses for optimal allocation of the PV-DG and the D-STATCOM have been determined using the sensitivity analysis. The assessment of integration of the PV-DGs and the D-STATCOMs has been studied for several objective functions, including the power losses, the voltage profile improvement, stability enhancement, and the total annual cost, which is comprised of the installation and operation cost of the PV-DGs and the D-STATCOMs, the purchased power from the grid, and the cost of the power loss. The effectiveness of the PV-DGs and the D-STATCOMs and the validity of the proposed algorithms have been tested and verified on different standard IEEE test systems such as IEEE-33, IEEE-69, IEEE-94, and IEEE-118 bus test systems and real test systems such as the 30-bus of the East Delta Network (EDN) and the 94-bus practical distribution system situated in Portugal.



A. M. A. El-Hameid et al., *Enhancement of Grid-Connected Photovoltaic Systems Using Artificial Intelligence*, https://doi.org/10.1007/978-3-031-29692-5_6

The allocation problem has been solved under deterministic and probabilistic conditions, i.e., considering the variations and uncertainties of the load demands and seasonal solar irradiances. From the study, the following conclusions are drawn:

- The developed algorithms are effective in interpreting the allocation of PV-DGs and the D-STATCOMs.
- In the case of optimal integration, a single PV-DG and the D-STATCOMs in IEEE-118 are at 20%, 30%, and 40% penetration levels; the annual cost reductions are 3.4%, 4.47%, and 5.72%, respectively, when compared to the base case. In the case of optimal integration of two hybrid PV-DGs and D-STATCOMs, the reductions in annual costs are 4.12%, 5.55%, and 6.69% under 20%, 30%, and 40% penetration levels, respectively.
- In the case of optimal integration of a single PV-DG and the D-STATCOMs in IEEE-94 under deterministic conditions, the cost is decreased by 14.88%, the total voltage deviation (TVD) is reduced by 73.36%, and the total voltage stability index (TVSI) is enhanced by 21.71% compared to the base case. In the case of including two hybrid systems, the reduction in cost and TVD are 15.68% and 77.96%, respectively, and TVSI is enhanced by 24.016%.
- In the case of optimal integration of a single PV-DG and the D-STATCOMs in IEEE-94 under probabilistic conditions, the cost is reduced by 10.917%, the time is minimized by 72.53%, and the enhancement is enhanced by 24.016% compared to the base case, while in the case of including two hybrid systems, the reduction in cost is 17.16% and 86.64%, respectively, and the enhancement is enhanced by 34.96%.
- Under the uncertainty condition, the PV-DGs and D-STATCOMs in IEEE-69 reduce total annual cost by 7.59% and 51.35%, respectively, in the case of optimal integration. Furthermore, voltage stability is improved by 5.05% when compared to the base case, and the optimal incorporation of two hybrid PV-DG and D-STATCOM can reduce total annual cost and voltage deviations by 10.78% and 57.55%, respectively. Also, the voltage stability is enhanced by 5.82% compared with the base case.
- In the case of optimal integration, the PV-DGs and the D-STATCOMs in IEEE-118 under the uncertainty condition of the expected cost, the voltage deviations are reduced by 8.25% and 27.34%, respectively. Furthermore, voltage stability is improved by 4.6% when compared to the base case, and the optimal incorporation of two hybrid PV-DGs and D-STATCOMs can reduce total annual cost and voltage deviations by 9.151% and 40.33%, respectively. Also, the voltage stability is enhanced by 6.57% compared with the base case.

# Appendix

## Appendix A

For 33-bus RDS, the active and reactive load demands are given in Table A.1 and the line data and base data of the system.



A. M. A. El-Hameid et al., *Enhancement of Grid-Connected Photovoltaic Systems Using Artificial Intelligence*, https://doi.org/10.1007/978-3-031-29692-5

**Table A.1** Bus data for 33-bus test system

| S. No | From bus | To bus | $R$ (Ω) | $X$ (Ω) | $P_L$ (KW) | $Q_L$ (KW) | $P_G$ | $Q_G$ |
|---|---|---|---|---|---|---|---|---|
| 0 | 0 | 0 | 0 | 0 | 0 | 0 | 0 | 0 |
| 1 | 1 | 2 | 0.0922 | 0.0477 | 100 | 60 | 0 | 0 |
| 2 | 2 | 3 | 0.493 | 0.2511 | 90 | 40 | 0 | 0 |
| 3 | 3 | 4 | 0.366 | 0.1864 | 120 | 80 | 0 | 0 |
| 4 | 4 | 5 | 0.3811 | 0.1941 | 60 | 30 | 0 | 0 |
| 5 | 5 | 6 | 0.819 | 0.707 | 60 | 20 | 0 | 0 |
| 6 | 6 | 7 | 0.1872 | 0.6188 | 200 | 100 | 0 | 0 |
| 7 | 7 | 8 | 1.7114 | 1.2351 | 200 | 100 | 0 | 0 |
| 8 | 8 | 9 | 1.03 | 0.74 | 60 | 20 | 0 | 0 |
| 9 | 9 | 10 | 1.04 | 0.74 | 60 | 20 | 0 | 0 |
| 10 | 10 | 11 | 0.1966 | 0.065 | 45 | 30 | 0 | 0 |
| 11 | 11 | 12 | 0.3744 | 0.1238 | 60 | 35 | 0 | 0 |
| 12 | 12 | 13 | 1.468 | 1.155 | 60 | 35 | 0 | 0 |
| 13 | 13 | 14 | 0.5416 | 0.7129 | 120 | 80 | 0 | 0 |
| 14 | 14 | 15 | 0.591 | 0.526 | 60 | 10 | 0 | 0 |
| 15 | 15 | 16 | 0.7463 | 0.545 | 60 | 20 | 0 | 0 |
| 16 | 16 | 17 | 1.289 | 1.721 | 60 | 20 | 0 | 0 |
| 17 | 17 | 18 | 0.732 | 0.574 | 90 | 40 | 0 | 0 |
| 18 | 2 | 19 | 0.164 | 0.1565 | 90 | 40 | 0 | 0 |
| 19 | 19 | 20 | 1.5042 | 1.3554 | 90 | 40 | 0 | 0 |
| 20 | 20 | 21 | 0.4095 | 0.4784 | 90 | 40 | 0 | 0 |
| 21 | 21 | 22 | 0.7089 | 0.9373 | 90 | 40 | 0 | 0 |
| 22 | 3 | 23 | 0.4512 | 0.3083 | 90 | 50 | 0 | 0 |
| 23 | 23 | 24 | 0.898 | 0.7091 | 420 | 200 | 0 | 0 |
| 24 | 24 | 25 | 0.896 | 0.7011 | 420 | 200 | 0 | 0 |
| 25 | 6 | 26 | 0.203 | 0.1034 | 60 | 25 | 0 | 0 |
| 26 | 26 | 27 | 0.2842 | 0.1447 | 60 | 25 | 0 | 0 |
| 27 | 27 | 28 | 1.059 | 0.9337 | 60 | 20 | 0 | 0 |
| 28 | 28 | 29 | 0.8042 | 0.7006 | 120 | 70 | 0 | 0 |
| 29 | 29 | 30 | 0.5075 | 0.2585 | 200 | 600 | 0 | 0 |
| 30 | 30 | 31 | 0.9744 | 0.963 | 150 | 70 | 0 | 0 |
| 31 | 31 | 32 | 0.3105 | 0.3619 | 210 | 100 | 0 | 0 |
| 32 | 32 | 33 | 0.341 | 0.5302 | 60 | 40 | 0 | 0 |

# Appendix B

For 30-bus RDS, the active and reactive load demands are given in Table B.1 and the line data and bus data of the system.

**Table B.1** Bus data for 30-bus test system

| S. No | From bus | To bus | $R$ (Ω) | $X$ (Ω) | $P_L$ (kW) | $Q_L$ (kVar) | $P_g$ (kW) | $Q_g$ (kVar) |
|---|---|---|---|---|---|---|---|---|
| 1 | 1 | 2 | 0.05630 | 0.031500 | 2875 | 1814 | 0 | 0 |
| 2 | 2 | 3 | 0.07155 | 0.025974 | 1100 | 695 | 0 | 0 |
| 3 | 3 | 4 | 0.01855 | 0.006734 | 1058 | 669 | 0 | 0 |
| 4 | 4 | 5 | 0.05565 | 0.020202 | 899 | 568 | 0 | 0 |
| 5 | 5 | 6 | 0.05300 | 0.019240 | 770 | 486 | 0 | 0 |
| 6 | 6 | 7 | 0.05300 | 0.019240 | 668 | 423 | 0 | 0 |
| 7 | 7 | 8 | 0.02120 | 0.007696 | 598 | 378 | 0 | 0 |
| 8 | 8 | 9 | 0.10070 | 0.036556 | 546 | 345 | 0 | 0 |
| 9 | 9 | 10 | 0.04505 | 0.016354 | 380 | 240 | 0 | 0 |
| 10 | 10 | 11 | 0.03975 | 0.014430 | 210 | 132 | 0 | 0 |
| 11 | 11 | 12 | 0.11130 | 0.040404 | 94.586 | 59.368 | 0 | 0 |
| 12 | 12 | 13 | 0.01325 | 0.004810 | 34.423 | 21.518 | 0 | 0 |
| 13 | 2 | 14 | 0.06360 | 0.023088 | 1772 | 1118 | 0 | 0 |
| 14 | 14 | 15 | 0.07155 | 0.025974 | 1640 | 1035 | 0 | 0 |
| 15 | 15 | 16 | 0.02650 | 0.009620 | 1452 | 915 | 0 | 0 |
| 16 | 16 | 17 | 0.01060 | 0.003848 | 1434 | 904 | 0 | 0 |
| 17 | 17 | 18 | 0.09275 | 0.033670 | 1212 | 765 | 0 | 0 |
| 18 | 2 | 19 | 0.01060 | 0.003848 | 1086 | 685 | 0 | 0 |
| 19 | 19 | 20 | 0.02650 | 0.009620 | 953 | 602 | 0 | 0 |
| 20 | 20 | 21 | 0.04505 | 0.016354 | 827 | 521 | 0 | 0 |
| 21 | 21 | 22 | 0.05300 | 0.019240 | 716 | 452 | 0 | 0 |
| 22 | 3 | 23 | 0.05300 | 0.019240 | 550 | 347 | 0 | 0 |
| 23 | 23 | 24 | 0.06630 | 0.024000 | 434 | 273 | 0 | 0 |
| 24 | 24 | 25 | 0.22530 | 0.081800 | 346 | 218 | 0 | 0 |
| 25 | 6 | 26 | 0.02650 | 0.009600 | 316 | 199 | 0 | 0 |
| 26 | 26 | 27 | 0.02650 | 0.009600 | 184 | 116 | 0 | 0 |
| 27 | 27 | 28 | 0.01330 | 0.004800 | 139 | 87.911 | 0 | 0 |
| 28 | 28 | 29 | 0.17230 | 0.062500 | 113 | 71.734 | 0 | 0 |
| 29 | 29 | 30 | 0.00800 | 0.002900 | 34.25 | 21.734 | 0 | 0 |

# Appendix C

For 118-bus RDS, the active and reactive load demands are given in Table C.1 and the line data and bus data of the system.

**Table C.1** Bus data for 118-bus test system

| S. No | From bus | To bus | *R* Ω | *X* (Ω) | $\mathbf{P}_L$ (kW) | $\mathbf{Q}_L$ (kVar) | $\mathbf{P}_g$ (kW) | $\mathbf{Q}_g$ (kVar) |
|---|---|---|---|---|---|---|---|---|
| 1 | 1 | 2 | 0.036 | 0.013 | 133.84 | 101.14 | 0 | 0 |
| 2 | 2 | 3 | 0.033 | 0.0119 | 16.214 | 11.292 | 0 | 0 |
| 3 | 2 | 4 | 0.045 | 0.0162 | 34.315 | 21.845 | 0 | 0 |
| 4 | 4 | 5 | 0.015 | 0.054 | 73.016 | 63.602 | 0 | 0 |
| 5 | 5 | 6 | 0.015 | 0.054 | 144.2 | 68.604 | 0 | 0 |
| 6 | 6 | 7 | 0.015 | 0.0125 | 104.47 | 61.725 | 0 | 0 |
| 7 | 7 | 8 | 0.018 | 0.014 | 28.547 | 11.503 | 0 | 0 |
| 8 | 8 | 9 | 0.021 | 0.063 | 87.56 | 51.073 | 0 | 0 |
| 9 | 2 | 10 | 0.166 | 0.1344 | 198.2 | 106.77 | 0 | 0 |
| 10 | 10 | 11 | 0.112 | 0.0789 | 146.8 | 75.995 | 0 | 0 |
| 11 | 11 | 12 | 0.187 | 0.313 | 26.04 | 18.687 | 0 | 0 |
| 12 | 12 | 13 | 0.142 | 0.1512 | 52.1 | 23.22 | 0 | 0 |
| 13 | 13 | 14 | 0.18 | 0.118 | 141.9 | 117.5 | 0 | 0 |
| 14 | 14 | 15 | 0.15 | 0.045 | 21.87 | 28.79 | 0 | 0 |
| 15 | 15 | 16 | 0.16 | 0.18 | 33.37 | 26.45 | 0 | 0 |
| 16 | 16 | 17 | 0.157 | 0.171 | 32.43 | 25.23 | 0 | 0 |
| 17 | 11 | 18 | 0.218 | 0.285 | 20.234 | 11.906 | 0 | 0 |
| 18 | 18 | 19 | 0.118 | 0.185 | 156.94 | 78.523 | 0 | 0 |
| 19 | 19 | 20 | 0.16 | 0.196 | 546.29 | 351.4 | 0 | 0 |
| 20 | 20 | 21 | 0.12 | 0.189 | 180.31 | 164.2 | 0 | 0 |
| 21 | 21 | 22 | 0.12 | 0.0789 | 93.167 | 54.594 | 0 | 0 |
| 22 | 22 | 23 | 1.41 | 0.723 | 85.18 | 39.65 | 0 | 0 |
| 23 | 23 | 24 | 0.293 | 0.1348 | 168.1 | 95.178 | 0 | 0 |
| 24 | 24 | 25 | 0.133 | 0.104 | 125.11 | 150.22 | 0 | 0 |
| 25 | 25 | 26 | 0.178 | 0.134 | 16.03 | 24.62 | 0 | 0 |
| 26 | 26 | 27 | 0.178 | 0.134 | 26.03 | 24.62 | 0 | 0 |
| 27 | 4 | 28 | 0.015 | 0.0296 | 594.56 | 522.62 | 0 | 0 |
| 28 | 28 | 29 | 0.012 | 0.0276 | 120.62 | 59.117 | 0 | 0 |
| 29 | 29 | 30 | 0.12 | 0.2766 | 102.38 | 99.554 | 0 | 0 |
| 30 | 30 | 31 | 0.21 | 0.243 | 513.4 | 318.5 | 0 | 0 |
| 31 | 31 | 32 | 0.12 | 0.054 | 475.25 | 456.14 | 0 | 0 |
| 32 | 32 | 33 | 0.178 | 0.234 | 151.43 | 136.79 | 0 | 0 |
| 33 | 33 | 34 | 0.178 | 0.234 | 205.38 | 83.302 | 0 | 0 |
| 34 | 34 | 35 | 0.154 | 0.162 | 131.6 | 93.082 | 0 | 0 |
| 35 | 30 | 36 | 0.187 | 0.261 | 448.4 | 369.79 | 0 | 0 |
| 36 | 36 | 37 | 0.133 | 0.099 | 440.52 | 321.64 | 0 | 0 |
| 37 | 29 | 38 | 0.33 | 0.194 | 112.54 | 55.134 | 0 | 0 |

(continued)

**Table C.1** (continued)

| S. No | From bus | To bus | $R\ \Omega$ | $X\ (\Omega)$ | $\mathbf{P_L}$ (kW) | $\mathbf{Q_L}$ (kVar) | $\mathbf{P_g}$ (kW) | $\mathbf{Q_g}$ (kVar) |
|---|---|---|---|---|---|---|---|---|
| 38 | 38 | 39 | 0.31 | 0.194 | 53.963 | 38.998 | 0 | 0 |
| 39 | 39 | 40 | 0.13 | 0.194 | 393.05 | 342.6 | 0 | 0 |
| 40 | 40 | 41 | 0.28 | 0.15 | 326.74 | 278.56 | 0 | 0 |
| 41 | 41 | 42 | 1.18 | 0.85 | 536.26 | 240.24 | 0 | 0 |
| 42 | 42 | 43 | 0.42 | 0.2436 | 76.247 | 66.562 | 0 | 0 |
| 43 | 43 | 44 | 0.27 | 0.0972 | 53.52 | 39.76 | 0 | 0 |
| 44 | 44 | 45 | 0.339 | 0.1221 | 40.328 | 31.964 | 0 | 0 |
| 45 | 45 | 46 | 0.27 | 0.1779 | 39.653 | 20.758 | 0 | 0 |
| 46 | 35 | 47 | 0.21 | 0.1383 | 66.195 | 42.361 | 0 | 0 |
| 47 | 47 | 48 | 0.12 | 0.0789 | 73.904 | 51.653 | 0 | 0 |
| 48 | 48 | 49 | 0.15 | 0.0987 | 114.77 | 57.965 | 0 | 0 |
| 49 | 49 | 50 | 0.15 | 0.0987 | 918.37 | 1205.1 | 0 | 0 |
| 50 | 50 | 51 | 0.24 | 0.1581 | 210.3 | 146.66 | 0 | 0 |
| 51 | 51 | 52 | 0.12 | 0.0789 | 66.68 | 56.608 | 0 | 0 |
| 52 | 52 | 53 | 0.405 | 0.1458 | 42.207 | 40.184 | 0 | 0 |
| 53 | 53 | 54 | 0.405 | 0.1458 | 433.74 | 283.41 | 0 | 0 |
| 54 | 29 | 55 | 0.391 | 0.141 | 62.1 | 26.86 | 0 | 0 |
| 55 | 55 | 56 | 0.406 | 0.1461 | 92.46 | 88.38 | 0 | 0 |
| 56 | 56 | 57 | 0.406 | 0.1461 | 85.188 | 55.436 | 0 | 0 |
| 57 | 57 | 58 | 0.706 | 0.5461 | 345.3 | 332.4 | 0 | 0 |
| 58 | 58 | 59 | 0.338 | 0.1218 | 22.5 | 16.83 | 0 | 0 |
| 59 | 59 | 60 | 0.338 | 0.1218 | 80.551 | 49.156 | 0 | 0 |
| 60 | 60 | 61 | 0.207 | 0.0747 | 95.86 | 90.758 | 0 | 0 |
| 61 | 61 | 62 | 0.247 | 0.8922 | 62.92 | 47.7 | 0 | 0 |
| 62 | 1 | 63 | 0.028 | 0.0418 | 478.8 | 463.74 | 0 | 0 |
| 63 | 63 | 64 | 0.117 | 0.2016 | 120.94 | 52.006 | 0 | 0 |
| 64 | 64 | 65 | 0.255 | 0.0918 | 139.11 | 100.34 | 0 | 0 |
| 65 | 65 | 66 | 0.21 | 0.0759 | 391.78 | 193.5 | 0 | 0 |
| 66 | 66 | 67 | 0.383 | 0.138 | 27.741 | 26.713 | 0 | 0 |
| 67 | 67 | 68 | 0.504 | 0.3303 | 52.814 | 25.257 | 0 | 0 |
| 68 | 68 | 69 | 0.406 | 0.1461 | 66.89 | 38.713 | 0 | 0 |
| 69 | 69 | 70 | 0.962 | 0.761 | 467.5 | 395.14 | 0 | 0 |
| 70 | 70 | 71 | 0.165 | 0.06 | 594.85 | 239.74 | 0 | 0 |
| 71 | 71 | 72 | 0.303 | 0.1092 | 132.5 | 84.363 | 0 | 0 |
| 72 | 72 | 73 | 0.303 | 0.1092 | 52.699 | 22.482 | 0 | 0 |
| 73 | 73 | 74 | 0.206 | 0.144 | 869.79 | 614.775 | 0 | 0 |
| 74 | 74 | 75 | 0.233 | 0.084 | 31.349 | 29.817 | 0 | 0 |
| 75 | 75 | 76 | 0.591 | 0.1773 | 192.39 | 122.43 | 0 | 0 |
| 76 | 76 | 77 | 0.126 | 0.0453 | 65.75 | 45.37 | 0 | 0 |
| 77 | 64 | 78 | 0.559 | 0.3687 | 238.15 | 223.22 | 0 | 0 |
| 78 | 78 | 79 | 0.186 | 0.1227 | 294.55 | 162.47 | 0 | 0 |
| 79 | 79 | 80 | 0.186 | 0.1227 | 485.57 | 437.92 | 0 | 0 |

(continued)

**Table C.1** (continued)

| S. No | From bus | To bus | $R\ \Omega$ | $X\ (\Omega)$ | $\mathbf{P_L}$ (kW) | $\mathbf{Q_L}$ (kVar) | $\mathbf{P_g}$ (kW) | $\mathbf{Q_g}$ (kVar) |
|---|---|---|---|---|---|---|---|---|
| 80 | 80 | 81 | 0.26 | 0.139 | 243.53 | 183.03 | 0 | 0 |
| 81 | 81 | 82 | 0.154 | 0.148 | 243.53 | 183.03 | 0 | 0 |
| 82 | 82 | 83 | 0.23 | 0.128 | 134.25 | 119.29 | 0 | 0 |
| 83 | 83 | 84 | 0.252 | 0.106 | 22.71 | 27.96 | 0 | 0 |
| 84 | 84 | 85 | 0.18 | 0.148 | 49.513 | 26.515 | 0 | 0 |
| 85 | 79 | 86 | 0.16 | 0.182 | 383.78 | 257.16 | 0 | 0 |
| 86 | 86 | 87 | 0.2 | 0.23 | 49.64 | 20.6 | 0 | 0 |
| 87 | 87 | 88 | 0.16 | 0.393 | 22.473 | 11.806 | 0 | 0 |
| 88 | 65 | 89 | 0.669 | 0.2412 | 62.93 | 42.96 | 0 | 0 |
| 89 | 89 | 90 | 0.266 | 0.1227 | 30.67 | 34.93 | 0 | 0 |
| 90 | 90 | 91 | 0.266 | 0.1227 | 62.53 | 66.79 | 0 | 0 |
| 91 | 91 | 92 | 0.266 | 0.1227 | 114.57 | 81.748 | 0 | 0 |
| 92 | 92 | 93 | 0.266 | 0.1227 | 81.292 | 66.526 | 0 | 0 |
| 93 | 93 | 94 | 0.233 | 0.115 | 31.733 | 15.96 | 0 | 0 |
| 94 | 94 | 95 | 0.496 | 0.138 | 33.32 | 60.48 | 0 | 0 |
| 95 | 91 | 96 | 0.196 | 0.18 | 531.28 | 224.85 | 0 | 0 |
| 96 | 96 | 97 | 0.196 | 0.18 | 507.03 | 367.42 | 0 | 0 |
| 97 | 97 | 98 | 0.1866 | 0.122 | 26.39 | 11.7 | 0 | 0 |
| 98 | 98 | 99 | 0.0746 | 0.318 | 45.99 | 30.392 | 0 | 0 |
| 99 | 1 | 100 | 0.0625 | 0.0265 | 100.66 | 47.572 | 0 | 0 |
| 100 | 100 | 101 | 0.1501 | 0.234 | 456.48 | 350.3 | 0 | 0 |
| 101 | 101 | 102 | 0.1347 | 0.0888 | 522.56 | 449.29 | 0 | 0 |
| 102 | 102 | 103 | 0.2307 | 0.1203 | 408.43 | 168.46 | 0 | 0 |
| 103 | 103 | 104 | 0.447 | 0.1608 | 141.48 | 134.25 | 0 | 0 |
| 104 | 104 | 105 | 0.1632 | 0.0588 | 104.43 | 66.024 | 0 | 0 |
| 105 | 105 | 106 | 0.33 | 0.099 | 96.793 | 83.647 | 0 | 0 |
| 106 | 106 | 107 | 0.156 | 0.0561 | 493.92 | 419.34 | 0 | 0 |
| 107 | 107 | 108 | 0.3819 | 0.1374 | 225.38 | 135.88 | 0 | 0 |
| 108 | 108 | 109 | 0.1626 | 0.0585 | 509.21 | 387.21 | 0 | 0 |
| 109 | 109 | 110 | 0.3819 | 0.1374 | 188.5 | 173.46 | 0 | 0 |
| 110 | 110 | 111 | 0.2445 | 0.0879 | 918.03 | 898.55 | 0 | 0 |
| 111 | 110 | 112 | 0.2088 | 0.0753 | 305.08 | 215.37 | 0 | 0 |
| 112 | 112 | 113 | 0.2301 | 0.0828 | 54.38 | 40.97 | 0 | 0 |
| 113 | 100 | 114 | 0.6102 | 0.2196 | 211.14 | 192.9 | 0 | 0 |
| 114 | 114 | 115 | 0.1866 | 0.127 | 67.009 | 53.336 | 0 | 0 |
| 115 | 115 | 116 | 0.3732 | 0.246 | 162.07 | 90.321 | 0 | 0 |
| 116 | 116 | 117 | 0.405 | 0.367 | 48.785 | 29.156 | 0 | 0 |
| 117 | 117 | 118 | 0.489 | 0.438 | 33.9 | 18.98 | 0 | 0 |

# Appendix D

For 94-bus RDS, the active and reactive load demands are given in Table D.1 and the line data and bus data of the system.

**Table D.1** Bus data for 94-bus test system

| S. No | From bus | To bus | $R$ (Ω) | $X$ (Ω) | $P_L$ (kW) | $Q_L$ (kVar) | $P_g$ (kW) | $Q_g$ (kVar) |
|---|---|---|---|---|---|---|---|---|
| 0 | 0 | 0 | 0 | 0 | 0 | 0 | 0 | 0 |
| 1 | 1 | 2 | 0.112 | 0.1873 | 22.5 | 10.9 | 0 | 0 |
| 2 | 2 | 3 | 0.0763 | 0.1274 | 240.3 | 116.4 | 0 | 0 |
| 3 | 3 | 4 | 0.1891 | 0.3161 | 24.3 | 11.8 | 0 | 0 |
| 4 | 4 | 5 | 0.2243 | 0.3749 | 0 | 0 | 0 | 0 |
| 5 | 5 | 6 | 0.2571 | 0.4297 | 0 | 0 | 0 | 0 |
| 6 | 6 | 7 | 0.134 | 0.2239 | 28.8 | 14 | 0 | 0 |
| 7 | 7 | 8 | 0.2986 | 0.4991 | 0 | 0 | 0 | 0 |
| 8 | 8 | 9 | 0.1953 | 0.3265 | 0 | 0 | 0 | 0 |
| 9 | 9 | 10 | 0.5097 | 0.8519 | 0 | 0 | 0 | 0 |
| 10 | 10 | 11 | 1.5303 | 1.5101 | 0 | 0 | 0 | 0 |
| 11 | 11 | 12 | 0.1889 | 0.1864 | 0 | 0 | 0 | 0 |
| 12 | 12 | 13 | 0.1816 | 0.1793 | 0 | 0 | 0 | 0 |
| 13 | 13 | 14 | 0.0661 | 0.0653 | 0 | 0 | 0 | 0 |
| 14 | 14 | 15 | 0.4115 | 0.4061 | 57.6 | 27.9 | 0 | 0 |
| 15 | 15 | 16 | 0.2584 | 0.255 | 0 | 0 | 0 | 0 |
| 16 | 16 | 17 | 0.2033 | 0.2006 | 0 | 0 | 0 | 0 |
| 17 | 17 | 18 | 0.7243 | 0.7148 | 18.9 | 9.2 | 0 | 0 |
| 18 | 18 | 19 | 0.2162 | 0.2134 | 0 | 0 | 0 | 0 |
| 19 | 19 | 20 | 0.35 | 0.3454 | 0 | 0 | 0 | 0 |
| 20 | 20 | 21 | 1.4775 | 0.3891 | 55.8 | 27 | 0 | 0 |
| 21 | 21 | 22 | 0.45 | 0.1185 | 40.5 | 19.6 | 0 | 0 |
| 22 | 22 | 23 | 0.771 | 0.203 | 54 | 26.2 | 0 | 0 |
| 23 | 23 | 24 | 0.885 | 0.2331 | 0 | 0 | 0 | 0 |
| 24 | 24 | 25 | 0.9915 | 0.2611 | 0 | 0 | 0 | 0 |
| 25 | 25 | 26 | 0.384 | 0.1011 | 46.8 | 22.7 | 0 | 0 |
| 26 | 26 | 27 | 0.7245 | 0.1908 | 0 | 0 | 0 | 0 |
| 27 | 27 | 28 | 1.185 | 0.3121 | 0 | 0 | 0 | 0 |
| 28 | 28 | 29 | 1.2353 | 0.6899 | 13.5 | 6.5 | 0 | 0 |
| 29 | 29 | 30 | 0.3557 | 0.1987 | 3.6 | 1.7 | 0 | 0 |
| 30 | 30 | 31 | 0.9494 | 0.3406 | 18 | 8.7 | 0 | 0 |
| 31 | 31 | 32 | 0.6899 | 0.3853 | 21.6 | 10.5 | 0 | 0 |
| 32 | 32 | 33 | 1.5707 | 0.8773 | 9 | 4.4 | 0 | 0 |
| 33 | 5 | 34 | 1.2655 | 0.454 | 64.8 | 31.4 | 0 | 0 |
| 34 | 5 | 35 | 0.1688 | 0.0943 | 65.7 | 31.8 | 0 | 0 |

(continued)

**Table D.1** (continued)

| S. No | From bus | To bus | $R$ (Ω) | $X$ (Ω) | $P_L$ (kW) | $Q_L$ (kVar) | $P_g$ (kW) | $Q_g$ (kVar) |
|---|---|---|---|---|---|---|---|---|
| 35 | 35 | 36 | 0.2741 | 0.1531 | 59.4 | 28.8 | 0 | 0 |
| 36 | 36 | 37 | 0.2552 | 0.1425 | 13.5 | 6.5 | 0 | 0 |
| 37 | 6 | 38 | 0.4165 | 0.2326 | 161.1 | 78 | 0 | 0 |
| 38 | 6 | 39 | 1.4835 | 0.3907 | 26.1 | 12.6 | 0 | 0 |
| 39 | 39 | 40 | 1.8 | 0.474 | 134.1 | 65 | 0 | 0 |
| 40 | 40 | 41 | 0.5177 | 0.2892 | 85.5 | 41.4 | 0 | 0 |
| 41 | 41 | 42 | 0.7148 | 0.3992 | 41.4 | 20.1 | 0 | 0 |
| 42 | 8 | 43 | 1.0575 | 0.2785 | 41.4 | 20.1 | 0 | 0 |
| 43 | 43 | 44 | 0.5198 | 0.2903 | 41.4 | 20.1 | 0 | 0 |
| 44 | 44 | 45 | 0.3341 | 0.1866 | 21.6 | 10.5 | 0 | 0 |
| 45 | 9 | 46 | 0.349 | 0.1949 | 25.2 | 12.2 | 0 | 0 |
| 46 | 10 | 47 | 0.5771 | 0.3223 | 45.9 | 22.2 | 0 | 0 |
| 47 | 47 | 48 | 0.3598 | 0.2009 | 36.9 | 17.9 | 0 | 0 |
| 48 | 48 | 49 | 0.7688 | 0.4294 | 63.9 | 31 | 0 | 0 |
| 49 | 49 | 50 | 0.2599 | 0.1451 | 68.4 | 33.1 | 0 | 0 |
| 50 | 50 | 51 | 0.8654 | 0.4833 | 27.9 | 13.5 | 0 | 0 |
| 51 | 10 | 52 | 0.5248 | 0.5179 | 81 | 39.2 | 0 | 0 |
| 52 | 52 | 53 | 0.1737 | 0.1714 | 69.3 | 33.6 | 0 | 0 |
| 53 | 53 | 54 | 0.6148 | 0.6068 | 62.1 | 30.1 | 0 | 0 |
| 54 | 54 | 55 | 0.198 | 0.1954 | 35.1 | 17 | 0 | 0 |
| 55 | 55 | 56 | 0.198 | 0.1954 | 205.2 | 99.4 | 0 | 0 |
| 56 | 56 | 57 | 0.285 | 0.2813 | 31.5 | 15.3 | 0 | 0 |
| 57 | 57 | 58 | 0.1429 | 0.141 | 521.1 | 252.4 | 0 | 0 |
| 58 | 58 | 59 | 0.3409 | 0.1904 | 212.4 | 102.9 | 0 | 0 |
| 59 | 59 | 60 | 0.3679 | 0.2055 | 39.6 | 19.2 | 0 | 0 |
| 60 | 60 | 61 | 0.3591 | 0.2006 | 45 | 21.8 | 0 | 0 |
| 61 | 61 | 62 | 0.3503 | 0.1957 | 17.1 | 8.3 | 0 | 0 |
| 62 | 62 | 63 | 0.4219 | 0.2356 | 21.6 | 10.5 | 0 | 0 |
| 63 | 63 | 64 | 1.538 | 0.5517 | 35.1 | 17 | 0 | 0 |
| 64 | 64 | 65 | 0.9788 | 0.3511 | 70.2 | 34 | 0 | 0 |
| 65 | 65 | 66 | 1.4911 | 0.5349 | 34.2 | 16.6 | 0 | 0 |
| 66 | 11 | 67 | 0.969 | 0.2552 | 22.5 | 10.9 | 0 | 0 |
| 67 | 67 | 68 | 0.6705 | 0.1766 | 45.9 | 22.2 | 0 | 0 |
| 68 | 12 | 69 | 0.4354 | 0.2432 | 33.3 | 16.1 | 0 | 0 |
| 69 | 13 | 70 | 0.4631 | 0.2586 | 36.9 | 17.9 | 0 | 0 |
| 70 | 70 | 71 | 0.2707 | 0.1512 | 45 | 21.8 | 0 | 0 |
| 71 | 15 | 72 | 0.6683 | 0.3732 | 75.6 | 36.6 | 0 | 0 |
| 72 | 72 | 73 | 0.8525 | 0.4762 | 67.5 | 32.7 | 0 | 0 |
| 73 | 16 | 74 | 0.3314 | 0.1851 | 27.9 | 13.5 | 0 | 0 |
| 74 | 18 | 75 | 0.405 | 0.2262 | 38.7 | 18.7 | 0 | 0 |
| 75 | 19 | 76 | 0.4367 | 0.2439 | 53.1 | 25.7 | 0 | 0 |

(continued)

**Table D.1** (continued)

| S. No | From bus | To bus | $R$ (Ω) | $X$ (Ω) | $P_L$ (kW) | $Q_L$ (kVar) | $P_g$ (kW) | $Q_g$ (kVar) |
|---|---|---|---|---|---|---|---|---|
| 76 | 19 | 77 | 0.3416 | 0.1908 | 65.7 | 31.8 | 0 | 0 |
| 77 | 77 | 78 | 0.2113 | 0.118 | 63 | 30.5 | 0 | 0 |
| 78 | 78 | 79 | 1.1249 | 0.4035 | 67.5 | 32.7 | 0 | 0 |
| 79 | 79 | 80 | 1.1738 | 0.6556 | 45 | 21.8 | 0 | 0 |
| 80 | 80 | 81 | 0.619 | 0.3457 | 9 | 4.4 | 0 | 0 |
| 81 | 81 | 82 | 0.5684 | 0.3174 | 16.2 | 7.8 | 0 | 0 |
| 82 | 20 | 83 | 0.8393 | 0.3011 | 67.5 | 32.7 | 0 | 0 |
| 83 | 83 | 84 | 0.2133 | 0.1191 | 296.1 | 143.4 | 0 | 0 |
| 84 | 84 | 85 | 0.3645 | 0.2036 | 72 | 34.9 | 0 | 0 |
| 85 | 85 | 86 | 0.3206 | 0.1791 | 76.5 | 37.1 | 0 | 0 |
| 86 | 22 | 87 | 0.7675 | 0.4286 | 90.9 | 44 | 0 | 0 |
| 87 | 24 | 88 | 1.5914 | 0.5709 | 72 | 34.9 | 0 | 0 |
| 88 | 25 | 89 | 0.702 | 0.3921 | 63 | 30.5 | 0 | 0 |
| 89 | 25 | 90 | 20.743 | 0.7441 | 21.6 | 10.5 | 0 | 0 |
| 90 | 90 | 91 | 0.678 | 0.2432 | 36.9 | 17.9 | 0 | 0 |
| 91 | 91 | 92 | 0.5738 | 0.3205 | 20.7 | 10 | 0 | 0 |
| 92 | 27 | 93 | 0.5913 | 0.3303 | 17.1 | 8.3 | 0 | 0 |
| 93 | 28 | 94 | 1.1865 | 0.3124 | 90 | 43.6 | 0 | 0 |

# Appendix E

For 69-bus RDS, the active and reactive load demands are given in Table E.1 and the line data and bus data of the system.

**Table E.1** Bus data for 69-bus test system

| S. No | From bus | To bus | $R$ (Ω) | $X$ (Ω) | $P_L$ (KW) | $Q_L$ (Kvar) | $P_G$ (KW) | $Q_G$ (Kvar) |
|---|---|---|---|---|---|---|---|---|
| 0 | 0 | 0 | 0 | 0 | 0 | 0 | 0 | 0 |
| 1 | 1 | 2 | 0.0005 | 0.0012 | 0 | 0 | 0 | 0 |
| 2 | 2 | 3 | 0.0005 | 0.0012 | 0 | 0 | 0 | 0 |
| 3 | 3 | 4 | 0.0015 | 0.0036 | 0 | 0 | 0 | 0 |
| 4 | 4 | 5 | 0.0251 | 0.0294 | 0 | 0 | 0 | 0 |
| 5 | 5 | 6 | 0.366 | 0.1864 | 2.6 | 2.2 | 0 | 0 |
| 6 | 6 | 7 | 0.3811 | 0.1941 | 40.4 | 30 | 0 | 0 |
| 7 | 7 | 8 | 0.0922 | 0.047 | 75 | 54 | 0 | 0 |
| 8 | 8 | 9 | 0.0493 | 0.0251 | 30 | 22 | 0 | 0 |
| 9 | 9 | 10 | 0.819 | 0.2707 | 28 | 19 | 0 | 0 |
| 10 | 10 | 11 | 0.1872 | 0.0691 | 145 | 104 | 0 | 0 |
| 11 | 11 | 12 | 0.7114 | 0.2351 | 145 | 104 | 0 | 0 |
| 12 | 12 | 13 | 1.03 | 0.34 | 8 | 5.5 | 0 | 0 |
| 13 | 13 | 14 | 1.044 | 0.345 | 8 | 5.5 | 0 | 0 |
| 14 | 14 | 15 | 1.058 | 0.3496 | 0 | 0 | 0 | 0 |
| 15 | 15 | 16 | 0.1966 | 0.065 | 45.5 | 30 | 0 | 0 |
| 16 | 16 | 17 | 0.3744 | 0.1238 | 60 | 35 | 0 | 0 |
| 17 | 17 | 18 | 0.0047 | 0.0016 | 60 | 35 | 0 | 0 |
| 18 | 18 | 19 | 0.3276 | 0.1083 | 0 | 0 | 0 | 0 |
| 19 | 19 | 20 | 0.2106 | 0.069 | 1 | 0.6 | 0 | 0 |
| 20 | 20 | 21 | 0.3416 | 0.1129 | 114 | 81 | 0 | 0 |
| 21 | 21 | 22 | 0.014 | 0.0046 | 5.3 | 3.5 | 0 | 0 |
| 22 | 22 | 23 | 0.1591 | 0.0526 | 0 | 0 | 0 | 0 |
| 23 | 23 | 24 | 0.3463 | 0.1145 | 28 | 20 | 0 | 0 |
| 24 | 24 | 25 | 0.7488 | 0.2745 | 0 | 0 | 0 | 0 |
| 25 | 25 | 26 | 0.3089 | 0.1021 | 14 | 10 | 0 | 0 |
| 26 | 26 | 27 | 0.1732 | 0.0572 | 14 | 10 | 0 | 0 |
| 27 | 3 | 28 | 0.0044 | 0.0108 | 26 | 18.6 | 0 | 0 |
| 28 | 28 | 29 | 0.064 | 0.1565 | 26 | 18.6 | 0 | 0 |
| 29 | 29 | 30 | 0.3978 | 0.1315 | 0 | 0 | 0 | 0 |
| 30 | 30 | 31 | 0.0702 | 0.0232 | 0 | 0 | 0 | 0 |
| 31 | 31 | 32 | 0.351 | 0.116 | 0 | 0 | 0 | 0 |
| 32 | 32 | 33 | 0.839 | 0.2816 | 14 | 10 | 0 | 0 |
| 33 | 33 | 34 | 1.708 | 0.5646 | 19.5 | 14 | 0 | 0 |
| 34 | 34 | 35 | 1.474 | 0.4673 | 6 | 4 | 0 | 0 |
| 35 | 3 | 36 | 0.0044 | 0.0108 | 26 | 18.55 | 0 | 0 |

(continued)

**Table E.1** (continued)

| S. No | From bus | To bus | $R$ (Ω) | $X$ (Ω) | $P_L$ (KW) | $Q_L$ (Kvar) | $P_G$ (KW) | $Q_G$ (Kvar) |
|---|---|---|---|---|---|---|---|---|
| 36 | 36 | 37 | 0.064 | 0.1565 | 26 | 18.55 | 0 | 0 |
| 37 | 37 | 38 | 0.1053 | 0.123 | 0 | 0 | 0 | 0 |
| 38 | 38 | 39 | 0.0304 | 0.0355 | 24 | 17 | 0 | 0 |
| 39 | 39 | 40 | 0.0018 | 0.0021 | 24 | 17 | 0 | 0 |
| 40 | 40 | 41 | 0.7283 | 0.8509 | 1.2 | 1 | 0 | 0 |
| 41 | 41 | 42 | 0.31 | 0.3623 | 0 | 0 | 0 | 0 |
| 42 | 42 | 43 | 0.041 | 0.0478 | 6 | 4.3 | 0 | 0 |
| 43 | 43 | 44 | 0.0092 | 0.0116 | 0 | 0 | 0 | 0 |
| 44 | 44 | 45 | 0.1089 | 0.1373 | 39.22 | 26.3 | 0 | 0 |
| 45 | 45 | 46 | 0.0009 | 0.0012 | 39.22 | 26.3 | 0 | 0 |
| 46 | 4 | 47 | 0.0034 | 0.0084 | 0 | 0 | 0 | 0 |
| 47 | 47 | 48 | 0.0851 | 0.2083 | 79 | 56.4 | 0 | 0 |
| 48 | 48 | 49 | 0.2898 | 0.7091 | 384.7 | 274.5 | 0 | 0 |
| 49 | 49 | 50 | 0.0822 | 0.2011 | 384 | 274.5 | 0 | 0 |
| 50 | 8 | 51 | 0.0928 | 0.0473 | 40.5 | 28.3 | 0 | 0 |
| 51 | 51 | 52 | 0.3319 | 0.1114 | 3.6 | 2.7 | 0 | 0 |
| 52 | 9 | 53 | 0.174 | 0.0886 | 4.35 | 3.5 | 0 | 0 |
| 53 | 53 | 54 | 0.203 | 0.1034 | 26.4 | 19 | 0 | 0 |
| 54 | 54 | 55 | 0.2842 | 0.1447 | 24 | 17.2 | 0 | 0 |
| 55 | 55 | 56 | 0.2813 | 0.1433 | 0 | 0 | 0 | 0 |
| 56 | 56 | 57 | 1.59 | 0.5337 | 0 | 0 | 0 | 0 |
| 57 | 57 | 58 | 0.7837 | 0.263 | 0 | 0 | 0 | 0 |
| 58 | 58 | 59 | 0.3042 | 0.1006 | 100 | 72 | 0 | 0 |
| 59 | 59 | 60 | 0.3861 | 0.1172 | 0 | 0 | 0 | 0 |
| 60 | 60 | 61 | 0.5075 | 0.2585 | 1244 | 888 | 0 | 0 |
| 61 | 61 | 62 | 0.0974 | 0.0496 | 32 | 23 | 0 | 0 |
| 62 | 62 | 63 | 0.145 | 0.0738 | 0 | 0 | 0 | 0 |
| 63 | 63 | 64 | 0.7105 | 0.3619 | 227 | 162 | 0 | 0 |
| 64 | 64 | 65 | 1.041 | 0.5302 | 59 | 42 | 0 | 0 |
| 65 | 11 | 66 | 0.2012 | 0.0611 | 18 | 13 | 0 | 0 |
| 66 | 66 | 67 | 0.0047 | 0.0014 | 18 | 13 | 0 | 0 |
| 67 | 12 | 68 | 0.7394 | 0.2444 | 28 | 20 | 0 | 0 |
| 68 | 68 | 69 | 0.0047 | 0.0016 | 28 | 20 | 0 | 0 |

# Appendix F

The data of seasonal hourly load profile are given in Table F.1

**Table F.1** The data of seasonal hourly load profile

| % Hour loading | Spring | Summer | Autonomy | Winter |
|---|---|---|---|---|
| 1 | 0.741 | 0.692 | 0.695 | 0.687 |
| 2 | 0.708 | 0.655 | 0.654 | 0.651 |
| 3 | 0.675 | 0.619 | 0.62 | 0.618 |
| 4 | 0.661 | 0.607 | 0.611 | 0.61 |
| 5 | 0.672 | 0.618 | 0.615 | 0.625 |
| 6 | 0.718 | 0.649 | 0.635 | 0.672 |
| 7 | 0.802 | 0.733 | 0.715 | 0.762 |
| 8 | 0.818 | 0.79 | 0.778 | 0.807 |
| 9 | 0.843 | 0.808 | 0.809 | 0.826 |
| 10 | 0.866 | 0.823 | 0.838 | 0.842 |
| 11 | 0.88 | 0.832 | 0.858 | 0.852 |
| 12 | 0.886 | 0.836 | 0.868 | 0.858 |
| 13 | 0.882 | 0.837 | 0.87 | 0.858 |
| 14 | 0.873 | 0.824 | 0.861 | 0.846 |
| 15 | 0.871 | 0.818 | 0.857 | 0.842 |
| 16 | 0.871 | 0.813 | 0.853 | 0.843 |
| 17 | 0.877 | 0.808 | 0.849 | 0.848 |
| 18 | 0.874 | 0.801 | 0.835 | 0.849 |
| 19 | 0.889 | 0.81 | 0.818 | 0.871 |
| 20 | 0.9 | 0.832 | 0.818 | 0.894 |
| 21 | 0.904 | 0.853 | 0.834 | 0.872 |
| 22 | 0.878 | 0.822 | 0.84 | 0.822 |
| 23 | 0.839 | 0.779 | 0.797 | 0.776 |
| 24 | 0.797 | 0.741 | 0.752 | 0.736 |

# Appendix G

```
%%%%%%%%%With 2PV 2 D-STATCOM%%%%%%%%%
%%%%%%%objective function%%%%%%%
format long

%%
Penetration_level = 1; % 50 of load level
%%
accuracy = 1e-6; maxiter = 4;
nbus = 118;
 busdata = [  1.0000         0     0     0     0
   2.0000    133.840  101.140     0     0
   3.0000     16.214   11.292     0     0
   4.0000     34.315   21.845     0     0
   5.0000     73.016   63.602     0     0
   6.0000    144.200   68.604     0     0
   7.0000    104.470   61.725     0     0
   8.0000     28.547   11.503     0     0
   9.0000     87.560   51.073     0     0
  10.0000    198.200  106.770     0     0
  11.0000    146.800   75.995     0     0
  12.0000     26.040   18.687     0     0
  13.0000     52.100   23.220     0     0
  14.0000    141.900  117.500     0     0
  15.0000     21.870   28.790     0     0
  16.0000     33.370   26.450     0     0
  17.0000     32.430   25.230     0     0
  18.0000     20.234   11.906     0     0
  19.0000    156.940   78.523     0     0
  20.0000    546.290  351.400     0     0
  21.0000    180.310  164.200     0     0
  22.0000     93.167   54.594     0     0
  23.0000     85.180   39.650     0     0
  24.0000    168.100   95.178     0     0
  25.0000    125.110  150.220     0     0
  26.0000     16.030   24.620     0     0
  27.0000     26.030   24.620     0     0
  28.0000    594.560  522.620     0     0
  29.0000    120.620   59.117     0     0
  30.0000    102.380   99.554     0     0
  31.0000    513.400  318.500     0     0
  32.0000    475.250  456.140     0     0
  33.0000    151.430  136.790     0     0
  34.0000    205.380   83.302     0     0
  35.0000    131.600   93.082     0     0
  36.0000    448.400  369.790     0     0
  37.0000    440.520  321.640     0     0
  38.0000    112.540   55.134     0     0
  39.0000     53.963   38.998     0     0
  40.0000    393.050  342.600     0     0
  41.0000    326.740  278.560     0     0
  42.0000    536.260  240.240     0     0
```

| | | | | |
|---|---|---|---|---|
| 43.0000 | 76.247 | 66.562 | 0 | 0 |
| 44.0000 | 53.520 | 39.760 | 0 | 0 |
| 45.0000 | 40.328 | 31.964 | 0 | 0 |
| 46.0000 | 39.653 | 20.758 | 0 | 0 |
| 47.0000 | 66.195 | 42.361 | 0 | 0 |
| 48.0000 | 73.904 | 51.653 | 0 | 0 |
| 49.0000 | 114.770 | 57.965 | 0 | 0 |
| 50.0000 | 918.370 | 1205.100 | 0 | 0 |
| 51.0000 | 210.300 | 146.660 | 0 | 0 |
| 52.0000 | 66.680 | 56.608 | 0 | 0 |
| 53.0000 | 42.207 | 40.184 | 0 | 0 |
| 54.0000 | 433.740 | 283.410 | 0 | 0 |
| 55.0000 | 62.100 | 26.860 | 0 | 0 |
| 56.0000 | 92.460 | 88.380 | 0 | 0 |
| 57.0000 | 85.188 | 55.436 | 0 | 0 |
| 58.0000 | 345.300 | 332.400 | 0 | 0 |
| 59.0000 | 22.500 | 16.830 | 0 | 0 |
| 60.0000 | 80.551 | 49.156 | 0 | 0 |
| 61.0000 | 95.860 | 90.758 | 0 | 0 |
| 62.0000 | 62.920 | 47.700 | 0 | 0 |
| 63.0000 | 478.800 | 463.740 | 0 | 0 |
| 64.0000 | 120.940 | 52.006 | 0 | 0 |
| 65.0000 | 139.110 | 100.340 | 0 | 0 |
| 66.0000 | 391.780 | 193.500 | 0 | 0 |
| 67.0000 | 27.741 | 26.713 | 0 | 0 |
| 68.0000 | 52.814 | 25.257 | 0 | 0 |
| 69.0000 | 66.890 | 38.713 | 0 | 0 |
| 70.0000 | 467.500 | 395.140 | 0 | 0 |
| 71.0000 | 594.850 | 239.740 | 0 | 0 |
| 72.0000 | 132.500 | 84.363 | 0 | 0 |
| 73.0000 | 52.699 | 22.482 | 0 | 0 |
| 74.0000 | 869.790 | 614.775 | 0 | 0 |
| 75.0000 | 31.349 | 29.817 | 0 | 0 |
| 76.0000 | 192.390 | 122.430 | 0 | 0 |
| 77.0000 | 65.750 | 45.370 | 0 | 0 |
| 78.0000 | 238.150 | 223.220 | 0 | 0 |
| 79.0000 | 294.550 | 162.470 | 0 | 0 |
| 80.0000 | 485.570 | 437.920 | 0 | 0 |
| 81.0000 | 243.530 | 183.030 | 0 | 0 |
| 82.0000 | 243.530 | 183.030 | 0 | 0 |
| 83.0000 | 134.250 | 119.290 | 0 | 0 |
| 84.0000 | 22.710 | 27.960 | 0 | 0 |
| 85.0000 | 49.513 | 26.515 | 0 | 0 |
| 86.0000 | 383.780 | 257.160 | 0 | 0 |
| 87.0000 | 49.640 | 20.600 | 0 | 0 |
| 88.0000 | 22.473 | 11.806 | 0 | 0 |
| 89.0000 | 62.930 | 42.960 | 0 | 0 |
| 90.0000 | 30.670 | 34.930 | 0 | 0 |
| 91.0000 | 62.530 | 66.790 | 0 | 0 |
| 92.0000 | 114.570 | 81.748 | 0 | 0 |
| 93.0000 | 81.292 | 66.526 | 0 | 0 |
| 94.0000 | 31.733 | 15.960 | 0 | 0 |
| 95.0000 | 33.320 | 60.480 | 0 | 0 |

```
 96.0000  531.280  224.850    0     0
 97.0000  507.030  367.420    0     0
 98.0000   26.390   11.700    0     0
 99.0000   45.990   30.392    0     0
100.0000  100.660   47.572    0     0
101.0000  456.480  350.300    0     0
102.0000  522.560  449.290    0     0
103.0000  408.430  168.460    0     0
104.0000  141.480  134.250    0     0
105.0000  104.430   66.024    0     0
106.0000   96.793   83.647    0     0
107.0000  493.920  419.340    0     0
108.0000  225.380  135.880    0     0
109.0000  509.210  387.210    0     0
110.0000  188.500  173.460    0     0
111.0000  918.030  898.550    0     0
112.0000  305.080  215.370    0     0
113.0000   54.380   40.970    0     0
114.0000  211.140  192.900    0     0
115.0000   67.009   53.336    0     0
116.0000  162.070   90.321    0     0
117.0000   48.785   29.156    0     0
118.0000   33.900   18.980    0     0];

linedata = [ 1.0000    1.0000    2.0000    0.0360    0.0130
  2.0000     2.0000    3.0000    0.0330    0.0119
  3.0000     2.0000    4.0000    0.0450    0.0162
  4.0000     4.0000    5.0000    0.0150    0.0540
  5.0000     5.0000    6.0000    0.0150    0.0540
  6.0000     6.0000    7.0000    0.0150    0.0125
  7.0000     7.0000    8.0000    0.0180    0.0140
  8.0000     8.0000    9.0000    0.0210    0.0630
  9.0000     2.0000   10.0000    0.1660    0.1344
 10.0000    10.0000   11.0000    0.1120    0.0789
 11.0000    11.0000   12.0000    0.1870    0.3130
 12.0000    12.0000   13.0000    0.1420    0.1512
 13.0000    13.0000   14.0000    0.1800    0.1180
 14.0000    14.0000   15.0000    0.1500    0.0450
 15.0000    15.0000   16.0000    0.1600    0.1800
 16.0000    16.0000   17.0000    0.1570    0.1710
 17.0000    11.0000   18.0000    0.2180    0.2850
 18.0000    18.0000   19.0000    0.1180    0.1850
 19.0000    19.0000   20.0000    0.1600    0.1960
 20.0000    20.0000   21.0000    0.1200    0.1890
 21.0000    21.0000   22.0000    0.1200    0.0789
 22.0000    22.0000   23.0000    1.4100    0.7230
 23.0000    23.0000   24.0000    0.2930    0.1348
 24.0000    24.0000   25.0000    0.1330    0.1040
 25.0000    25.0000   26.0000    0.1780    0.1340
 26.0000    26.0000   27.0000    0.1780    0.1340
 27.0000     4.0000   28.0000    0.0150    0.0296
 28.0000    28.0000   29.0000    0.0120    0.0276
 29.0000    29.0000   30.0000    0.1200    0.2766
```

| | | | | |
|---|---|---|---|---|
| 30.0000 | 30.0000 | 31.0000 | 0.2100 | 0.2430 |
| 31.0000 | 31.0000 | 32.0000 | 0.1200 | 0.0540 |
| 32.0000 | 32.0000 | 33.0000 | 0.1780 | 0.2340 |
| 33.0000 | 33.0000 | 34.0000 | 0.1780 | 0.2340 |
| 34.0000 | 34.0000 | 35.0000 | 0.1540 | 0.1620 |
| 35.0000 | 30.0000 | 36.0000 | 0.1870 | 0.2610 |
| 36.0000 | 36.0000 | 37.0000 | 0.1330 | 0.0990 |
| 37.0000 | 29.0000 | 38.0000 | 0.3300 | 0.1940 |
| 38.0000 | 38.0000 | 39.0000 | 0.3100 | 0.1940 |
| 39.0000 | 39.0000 | 40.0000 | 0.1300 | 0.1940 |
| 40.0000 | 40.0000 | 41.0000 | 0.2800 | 0.1500 |
| 41.0000 | 41.0000 | 42.0000 | 1.1800 | 0.8500 |
| 42.0000 | 42.0000 | 43.0000 | 0.4200 | 0.2436 |
| 43.0000 | 43.0000 | 44.0000 | 0.2700 | 0.0972 |
| 44.0000 | 44.0000 | 45.0000 | 0.3390 | 0.1221 |
| 45.0000 | 45.0000 | 46.0000 | 0.2700 | 0.1779 |
| 46.0000 | 35.0000 | 47.0000 | 0.2100 | 0.1383 |
| 47.0000 | 47.0000 | 48.0000 | 0.1200 | 0.0789 |
| 48.0000 | 48.0000 | 49.0000 | 0.1500 | 0.0987 |
| 49.0000 | 49.0000 | 50.0000 | 0.1500 | 0.0987 |
| 50.0000 | 50.0000 | 51.0000 | 0.2400 | 0.1581 |
| 51.0000 | 51.0000 | 52.0000 | 0.1200 | 0.0789 |
| 52.0000 | 52.0000 | 53.0000 | 0.4050 | 0.1458 |
| 53.0000 | 53.0000 | 54.0000 | 0.4050 | 0.1458 |
| 54.0000 | 29.0000 | 55.0000 | 0.3910 | 0.1410 |
| 55.0000 | 55.0000 | 56.0000 | 0.4060 | 0.1461 |
| 56.0000 | 56.0000 | 57.0000 | 0.4060 | 0.1461 |
| 57.0000 | 57.0000 | 58.0000 | 0.7060 | 0.5461 |
| 58.0000 | 58.0000 | 59.0000 | 0.3380 | 0.1218 |
| 59.0000 | 59.0000 | 60.0000 | 0.3380 | 0.1218 |
| 60.0000 | 60.0000 | 61.0000 | 0.2070 | 0.0747 |
| 61.0000 | 61.0000 | 62.0000 | 0.2470 | 0.8922 |
| 62.0000 | 1.0000 | 63.0000 | 0.0280 | 0.0418 |
| 63.0000 | 63.0000 | 64.0000 | 0.1170 | 0.2016 |
| 64.0000 | 64.0000 | 65.0000 | 0.2550 | 0.0918 |
| 65.0000 | 65.0000 | 66.0000 | 0.2100 | 0.0759 |
| 66.0000 | 66.0000 | 67.0000 | 0.3830 | 0.1380 |
| 67.0000 | 67.0000 | 68.0000 | 0.5040 | 0.3303 |
| 68.0000 | 68.0000 | 69.0000 | 0.4060 | 0.1461 |
| 69.0000 | 69.0000 | 70.0000 | 0.9620 | 0.7610 |
| 70.0000 | 70.0000 | 71.0000 | 0.1650 | 0.0600 |
| 71.0000 | 71.0000 | 72.0000 | 0.3030 | 0.1092 |
| 72.0000 | 72.0000 | 73.0000 | 0.3030 | 0.1092 |
| 73.0000 | 73.0000 | 74.0000 | 0.2060 | 0.1440 |
| 74.0000 | 74.0000 | 75.0000 | 0.2330 | 0.0840 |
| 75.0000 | 75.0000 | 76.0000 | 0.5910 | 0.1773 |
| 76.0000 | 76.0000 | 77.0000 | 0.1260 | 0.0453 |
| 77.0000 | 64.0000 | 78.0000 | 0.5590 | 0.3687 |
| 78.0000 | 78.0000 | 79.0000 | 0.1860 | 0.1227 |
| 79.0000 | 79.0000 | 80.0000 | 0.1860 | 0.1227 |
| 80.0000 | 80.0000 | 81.0000 | 0.2600 | 0.1390 |
| 81.0000 | 81.0000 | 82.0000 | 0.1540 | 0.1480 |
| 82.0000 | 82.0000 | 83.0000 | 0.2300 | 0.1280 |

```
 83.0000   83.0000    84.0000    0.2520    0.1060
 84.0000   84.0000    85.0000    0.1800    0.1480
 85.0000   79.0000    86.0000    0.1600    0.1820
 86.0000   86.0000    87.0000    0.2000    0.2300
 87.0000   87.0000    88.0000    0.1600    0.3930
 88.0000   65.0000    89.0000    0.6690    0.2412
 89.0000   89.0000    90.0000    0.2660    0.1227
 90.0000   90.0000    91.0000    0.2660    0.1227
 91.0000   91.0000    92.0000    0.2660    0.1227
 92.0000   92.0000    93.0000    0.2660    0.1227
 93.0000   93.0000    94.0000    0.2330    0.1150
 94.0000   94.0000    95.0000    0.4960    0.1380
 95.0000   91.0000    96.0000    0.1960    0.1800
 96.0000   96.0000    97.0000    0.1960    0.1800
 97.0000   97.0000    98.0000    0.1866    0.1220
 98.0000   98.0000    99.0000    0.0746    0.3180
 99.0000    1.0000   100.0000    0.0625    0.0265
100.0000  100.0000   101.0000    0.1501    0.2340
101.0000  101.0000   102.0000    0.1347    0.0888
102.0000  102.0000   103.0000    0.2307    0.1203
103.0000  103.0000   104.0000    0.4470    0.1608
104.0000  104.0000   105.0000    0.1632    0.0588
105.0000  105.0000   106.0000    0.3300    0.0990
106.0000  106.0000   107.0000    0.1560    0.0561
107.0000  107.0000   108.0000    0.3819    0.1374
108.0000  108.0000   109.0000    0.1626    0.0585
109.0000  109.0000   110.0000    0.3819    0.1374
110.0000  110.0000   111.0000    0.2445    0.0879
111.0000  110.0000   112.0000    0.2088    0.0753
112.0000  112.0000   113.0000    0.2301    0.0828
113.0000  100.0000   114.0000    0.6102    0.2196
114.0000  114.0000   115.0000    0.1866    0.1270
115.0000  115.0000   116.0000    0.3732    0.2460
116.0000  116.0000   117.0000    0.4050    0.3670
117.0000  117.0000   118.0000    0.4890    0.4380
];
      %%
    %hour    spring     summer    Atonomy    winter
Loading = [1.0000   0.7410    0.6920    0.6950    0.6870
   2.0000      0.7080    0.6550    0.6540    0.6510
   3.0000      0.6750    0.6190    0.6200    0.6180
   4.0000      0.6610    0.6070    0.6110    0.6100
   5.0000      0.6720    0.6180    0.6150    0.6250
   6.0000      0.7180    0.6490    0.6350    0.6720
   7.0000      0.8020    0.7330    0.7150    0.7620
   8.0000      0.8180    0.7900    0.7780    0.8070
   9.0000      0.8430    0.8080    0.8090    0.8260
  10.0000      0.8660    0.8230    0.8380    0.8420
  11.0000      0.8800    0.8320    0.8580    0.8520
  12.0000      0.8860    0.8360    0.8680    0.8580
  13.0000      0.8820    0.8370    0.8700    0.8580
  14.0000      0.8730    0.8240    0.8610    0.8460
  15.0000      0.8710    0.8180    0.8570    0.8420
```

```
16.0000      0.8710   0.8130   0.8530   0.8430
17.0000      0.8770   0.8080   0.8490   0.8480
18.0000      0.8740   0.8010   0.8350   0.8490
19.0000      0.8890   0.8100   0.8180   0.8710
20.0000      0.9000   0.8320   0.8180   0.8940
21.0000      0.9040   0.8530   0.8340   0.8720
22.0000      0.8780   0.8220   0.8400   0.8220
23.0000      0.8390   0.7790   0.7970   0.7760
24.0000      0.7970   0.7410   0.7520   0.7360];

%%      winter          spring       summer       autumn
rradiance =  [0       0       0       0
        0        0        0         0
        0        0        0         0
        0        0        0         0
        0        0        0         0
        0        0        0         0
        0   13.8690  28.3120   0.0020
   0.0060  150.8170 205.7910  37.8490
  39.6730  366.5050 418.7880 215.6190
 201.2860 553.9440 596.0590 397.8950
 361.8990 717.6510 757.6500 545.7280
 474.4810 825.2850 860.9530 636.3370
 516.1320 878.4380 881.4250 663.3410
 506.2450 852.4560 837.0310 626.1620
 425.1740 766.8170 770.8930 525.0770
 293.1390 622.7320 607.0090 368.3480
 108.5580 430.6630 468.4420 132.3930
   0.7270 189.0940 300.6030  12.5290
        0   32.5410 113.1220   0.4930
        0    0.0020   0.0730        0
        0        0        0         0
        0        0        0         0
        0        0        0         0
        0        0        0         0];
%%
Gstd=1000; % standard of solar
Rc=120;
Gs=rradiance ;
Gs_winter = Gs(:,1) ;
Gs_Spring = Gs(:,2) ;
Gs_Summer = Gs(:,3) ;
Gs_Atonomy = Gs(:,4);


%%
LL1 = round(100*x(1));
LL2 = round(100*x(2));
L1 = LL1;      % Location of frist PV1
L2 = LL2;      % Location of frist PV1
L3 = L1;     % Location of frist D-STATCOM
L4 = L2;    % Location of frist D-STATCOM
PSR1 = round((100*x(3)));
PSR2 = round((100*x(4)));
```

```
Q_D-STATCOM1 = round((100*x(5)));
Q_D-STATCOM2 = round((100*x(6)));

busdata(L3,5) = Q_D-STATCOM1;
busdata(L4,5) = Q_D-STATCOM2;
  %%
% %% DSTAT^COM cost
Q_STATCOM = Q_D-STATCOM1;
B = 0.1;
K_Statcom = 50;
N_Statcom = 30;
AA = (1+B)^N_Statcom;
Statcom_Cost = Q_STATCOM*K_Statcom * ((AA*B)/(AA-1));
%%
orginal_busdata = busdata;
winter_loading = Loading(:,2);
spring_loading = Loading(:,3);
summer_loading = Loading(:,4);
Atonomy_loading = Loading(:,5);
Total_PL = sum (orginal_busdata(:,2));
Total_QL = sum (orginal_busdata(:,3));
%%

%%
%%%%%% spring_loading %%%%%
% disp('%%%%%% spring_loading %%%%%')
for hh = 1:length(spring_loading)

 busdata(:,2)  =  spring_loading(hh)*orginal_busdata(:,2);
 busdata(:,3)  =  spring_loading(hh)*orginal_busdata(:,3);

 if  Gs_Spring(hh)   < Rc
   PS_Spring1(hh)=PSR1*(((Gs_Spring(hh))^2)/(Gstd*Rc));
   PS_Spring2(hh)=PSR1*(((Gs_Spring(hh))^2)/(Gstd*Rc));

elseif   Gs_Spring(hh) > Rc ||  Gs_Spring(hh)== Rc
   PS_Spring1(hh)= PSR1 *(Gs_Spring(hh)/Gstd);
   PS_Spring2(hh)= PSR2 *(Gs_Spring(hh)/Gstd);
elseif Gs_Spring(hh) > Gstd
  PS_Spring1(hh)=PSR1  ;
  PS_Spring2(hh)=PSR2  ;

 end

  busdata(L1,4) = PS_Spring1(hh);
  busdata(L2,4) = PS_Spring2(hh);

PerUnit ;
backward_forward_sweep ;
Calculation1 ;
Plloss_spring(hh) = TotalActiveLoss_KW;
VDD_spring(hh) = VDD;
```

```
VSI_spring(hh) = Sum_VSI;
Real_Power_Supplied_From_Substation_KW_spring(hh) =
Real_Power_Supplied_From_Substation_KW;

 %__________________ Constraints ____________________

for i = 1: length(V_mag_no1)
if V_mag_no1(i)> 1.05 || V_mag_no1(i)< 0.90
   k1(i) = 1000;
else
   k1(i) = 0;
end
end
k1 = sum(k1);
k11(hh) = sum(k1);

end
PS_Spring_out1 = sum (PS_Spring1)*91.25;
PS_Spring_out2 = sum (PS_Spring2)*91.25;

PS_Spring_out = PS_Spring_out1+PS_Spring_out2;

Plloss_spring;
Sum_Plloss_spring = sum(Plloss_spring) ;
E_spring = Sum_Plloss_spring *91.25 ;
Sum_VD_spring  = sum(VDD_spring) ;
Sum_VSI_spring = sum(VSI_spring) ;
E_Station_spring = sum
(Real_Power_Supplied_From_Substation_KW_spring )*91.25;

%%
%%%%%% summer_loading %%%%%
% disp('%%%%%% summer_loading %%%%%')
for hh = 1:length(summer_loading)
 hh ;
 busdata(:,2) = summer_loading(hh)*orginal_busdata(:,2);
 busdata(:,3) = summer_loading(hh)*orginal_busdata(:,3);

  if  Gs_Summer(hh)  < Rc
   PS_Summer1(hh)=PSR1*(((Gs_Summer(hh))^2)/(Gstd*Rc));
   PS_Summer2(hh)=PSR2*(((Gs_Summer(hh))^2)/(Gstd*Rc));

 elseif  Gs_Summer(hh)>=Rc  && Gs_Summer(hh)<= Gstd
  PS_Summer1(hh)=PSR1 *(Gs_Summer(hh)/Gstd);
  PS_Summer2(hh)=PSR2 *(Gs_Summer(hh)/Gstd);

  elseif  Gs_Summer(hh)> Gstd
  PS_Summer1(hh)=PSR1  ;
  PS_Summer2(hh)=PSR2  ;

  end
```

```
 busdata(L1,4) = PS_Summer1(hh);
 busdata(L2,4) = PS_Summer2(hh);

PerUnit
backward_forward_sweep
Calculation1
Plloss_summer(hh) = TotalActiveLoss_KW;
VDD_summer(hh) = VDD;
VSI_summer(hh) = Sum_VSI;
Real_Power_Supplied_From_Substation_KW_summer(hh) =
Real_Power_Supplied_From_Substation_KW;

%_________________ Constraints ___________________

for i = 1: length(V_mag_no1)
if V_mag_no1(i) > 1.05 || V_mag_no1(i) < 0.90
   k2(i) = 1000;
else
   k2(i) = 0;
end
end
k2 = sum(k2);
k22(hh) = sum(k2);
end
PS_Summer_out1 =  sum (PS_Summer1)*91.25;
PS_Summer_out2 =  sum (PS_Summer2)*91.25;
PS_Summer_out = PS_Summer_out1+PS_Summer_out2;

Plloss_summer;
Sum_Plloss_summer = sum(Plloss_summer);
E_summer = Sum_Plloss_summer *91.25 ;
Sum_VD_summer  = sum(VDD_summer);
Sum_VSI_summer = sum(VSI_summer);
E_Station_summer = sum
(Real_Power_Supplied_From_Substation_KW_summer)*91.25;

%%

%%%%%% Atonomy_loading %%%%%
% disp('%%%%%% Atonomy_loading %%%%%')
for hh = 1:length(Atonomy_loading)
 hh;
 busdata(:,2) = Atonomy_loading(hh)*orginal_busdata(:,2);
 busdata(:,3) = Atonomy_loading(hh)*orginal_busdata(:,3);

if   Gs_Atonomy(hh)   < Rc
  PS_Atonomy1(hh) = PSR1*(((Gs_Atonomy(hh))^2)/(Gstd*Rc));
  PS_Atonomy2(hh) = PSR2*(((Gs_Atonomy(hh))^2)/(Gstd*Rc));

 elseif  Gs_Atonomy(hh) >= Rc  &&  Gs_Atonomy(hh) <= Gstd
  PS_Atonomy1(hh) = PSR1 * (Gs_Atonomy(hh)/Gstd);
  PS_Atonomy2(hh) = PSR2 * (Gs_Atonomy(hh)/Gstd);
```

```
 elseif  Gs_Atonomy(hh)>  Gstd
 PS_Atonomy1(hh)=PSR1;
 PS_Atonomy2(hh)=PSR2;

 end
  busdata(L1,4) = PS_Atonomy1(hh);
  busdata(L2,4) = PS_Atonomy2(hh);

PerUnit
backward_forward_sweep
Calculation1
Plloss_Atonomy(hh) = TotalActiveLoss_KW;
VDD_Atonomy(hh) = VDD;
VSI_Atonomy(hh)  = Sum_VSI;
Real_Power_Supplied_From_Substation_KW_Atonomy(hh)  =
Real_Power_Supplied_From_Substation_KW;

 %_________________ Constraints ___________________

for i = 1: length(V_mag_no1)
if V_mag_no1(i)> 1.05 || V_mag_no1(i)< 0.90
   k3(i) = 1000;
else
   k3(i) = 0;
end
end
k3 = sum(k3);
k33(hh) = sum(k3);
end
PS_Atonomy_out1 = sum (PS_Atonomy1)*91.25;
PS_Atonomy_out2 = sum (PS_Atonomy2)*91.25;
PS_Atonomy_out  = PS_Atonomy_out1+PS_Atonomy_out2;

Plloss_Atonomy;
Sum_Plloss_Atonomy  = sum(Plloss_Atonomy );
E_Atonomy = Sum_Plloss_Atonomy *91.25 ;
Sum_VD_Atonomy   = sum(VDD_Atonomy);
Sum_VSI_Atonomy  = sum(VSI_Atonomy);
E_Station_Atonomy =  sum
(Real_Power_Supplied_From_Substation_KW_Atonomy)*91.25;

%%
%%%%%% winter_loading %%%%%
% disp('%%%%%% winter_loading %%%%%')
for hh = 1:length(winter_loading)
 hh ;
 busdata(:,2)  = winter_loading(hh)*orginal_busdata(:,2);
 busdata(:,3) = winter_loading(hh)*orginal_busdata(:,3);
if   Gs_winter(hh)  <  Rc
   PS_Winter1(hh) = PSR1*(((Gs_winter(hh))^2)/(Gstd*Rc));
   PS_Winter2(hh) = PSR2*(((Gs_winter(hh))^2)/(Gstd*Rc));
```

```
 elseif  Gs_winter(hh)>=Rc  && Gs_winter(hh)<= Gstd
   PS_Winter1(hh)= PSR1 *(Gs_winter(hh)/Gstd);
   PS_Winter2(hh)= PSR2 *(Gs_winter(hh)/Gstd);

  elseif Gs_winter(hh)> Gstd
  PS_Winter1(hh)=PSR1;
  PS_Winter2(hh)=PSR2;

 end

busdata(L1,4) = PS_Winter1(hh);
busdata(L2,4) = PS_Winter2(hh);

PerUnit
backward_forward_sweep
Calculation1
Plloss_winter(hh) = TotalActiveLoss_KW;
VDD_winter(hh) = VDD;
VSI_winter(hh)  = Sum_VSI;
Real_Power_Supplied_From_Substation_KW_winter(hh) =
Real_Power_Supplied_From_Substation_KW;

 %_________________ Constraints ___________________

for i = 1: length(V_mag_no1)
if V_mag_no1(i)> 1.05 || V_mag_no1(i)< 0.90
  k4(i) = 1000;
else
  k4(i) = 0;
end
end
k4 = sum(k4);
k44(hh) = sum(k4);
k44(hh);
end
PS_Winter_out1= sum (PS_Winter1)*91.25;
PS_Winter_out2= sum (PS_Winter2)*91.25;
PS_Winter_out  = PS_Winter_out1+PS_Winter_out2;

Plloss_winter;
Sum_Plloss_winter  = sum(Plloss_winter) ;
E_winter= Sum_Plloss_winter *91.25 ;
Sum_VD_winter   = sum(VDD_winter);
Sum_VSI_winter  = sum(VSI_winter);
E_Station_winter=  sum
(Real_Power_Supplied_From_Substation_KW_winter)*91.25;

%%
%%%%%%%%%%%%%%%%%%%%%%%%
 Power_PV =PSR1 + PSR2;
%  % Cost of PV
B = 0.1;
```

```
CSDG = 770;
N_PV = 20;
AA = (1+B)^N_PV;
Capital_cost = Power_PV*CSDG * ((AA*B)/(AA-1));

Total_KWh_PV = PS_Winter_out + PS_Atonomy_out + PS_Summer_out +
PS_Spring_out;
a = Capital_cost ;
O_M_cost  = 0.01;
Fuel_cost = 0;
b = O_M_cost +Fuel_cost;
Cost_PV = a+ b*Total_KWh_PV;

%%     result
%  Penteration level constraints
Pload = Total_PL*Penetration_level;

if Power_PV > 22709
  m5 = 1000000;
else
  m5 = 0;
end
if Q_STATCOM > 17000
  m6 = 1000000;
else
  m6 = 0;
end
%%    result
%%
Power_PV =PSR1+PSR2 ;
%  % Cost of PV
B = 0.1;
CSDG = 770;
N_PV = 20;
AA = (1+B)^N_PV;
Capital_cost = Power_PV*CSDG * ((AA*B)/(AA-1));

Total_KWh_PV = PS_Winter_out + PS_Atonomy_out + PS_Summer_out +
PS_Spring_out;
a = Capital_cost ;
O_M_cost  = 0.01;
Fuel_cost = 0;
b = O_M_cost +Fuel_cost;
Cost_PV = a+ b*Total_KWh_PV;
%%    result
% Penteration level constraints
Pload = Total_PL;
% *Penetration_level;

if Power_PV > 22709
  m5 = 1000000;
else
```

```
  m5 = 0;
end
if Q_STATCOM > 17000
  m6 = 1000000;
else
  m6 = 0;
end
%
%%
format long

Total_E_from_substation =  E_Station_spring + E_Station_summer...
                   + E_Station_Atonomy + E_Station_winter
Total_Enery_loss = E_spring + E_summer + E_Atonomy + E_winter

Cost_Total_Enery_loss = 0.06*Total_Enery_loss
Ke =  0.096;
Purchasing_power_from_Substation = Total_E_from_substation * Ke

Total_VD = 91.25*(Sum_VD_spring + Sum_VD_summer +Sum_VD_Atonomy +
Sum_VD_winter)

Total_VSI = 91.25*(Sum_VSI_spring  + Sum_VSI_summer +Sum_VSI_Atonomy
+Sum_VSI_winter)
Ke =  0.096;
m1= sum(k11)
m2=sum(k22)
m3=sum(k33)
m4=sum(k44)
m5
m6
All_over_cost = Cost_Total_Enery_loss +
Purchasing_power_from_Substation +Statcom_Cost +Cost_PV
Statcom_Cost
Cost_PV

%o = All_over_cost+ 1000*m1 + 1000*m2 +1000*m3 +1000*m4 +100000*m5+ m6;
optimal_location_hybrid1 =  L1
optimal_location_hybrid2 =  L2
optimal_location_D-STATCOM_Hybrid1  = L1
optimal_location_D-STATCOM_Hybrid2  = L2
Optimal_size_KW_PV1 = PSR1
Optimal_size_KW_PV2 = PSR2
optimal_size_D-STATCOM_hybrid2  = Q_D-STATCOM1
optimal_size_D-STATCOM_hybrid2  = Q_D-STATCOM2
Net_Saving = 1.205051854113218e+07-All_over_cost
O1 = All_over_cost/1.205051854113218e+07;
O2 = Total_VD/ 3.5605e+04;
O3 = 1/Total_VSI;

O3 = 1/Total_VSI;% O4= Total_Enery_loss/4.935812577976250e+05;
w1=0.5;
```

```
w2=0.25;
w3=0.25;
o = w1*O1+w2*O2+w3*O3+ 1000*m1 +1000*m2 +1000*m3 +10000*m4+1000*m5 +m6

disp('---------------------------------------')
All_over_cost
Total_VD
Total_VSI
disp('---------------------------------------')
%%
Pss1PV=[PS_Winter1' PS_Spring1' PS_Summer1' PS_Atonomy1']
 Pss2PV=[PS_Winter2' PS_Spring2' PS_Summer2' PS_Atonomy2']

% Power_grid2 = [ Real_Power_Supplied_From_Substation_KW_winter'
Real_Power_Supplied_From_Substation_KW_spring'
Real_Power_Supplied_From_Substation_KW_Atonomy'
Real_Power_Supplied_From_Substation_KW_summer']
 LossLoss_all = [ Plloss_winter' Plloss_spring' Plloss_summer'
Plloss_Atonomy']

%%
format long
fprintf('\n')
fprintf('==============================================
==============================')
fprintf('\n')
fprintf('Minimum voltage (p.u)                              :');
fprintf('%3.5f',MinVoltage);
fprintf(' @ bus ');
fprintf('%4.f',Min_bus);

fprintf('\n')
fprintf('Maximum voltage (p.u) excluding the slack bus  :');
fprintf('%4.5f',MaxVoltage);
fprintf(' @ bus ');
fprintf('%4.f',Max_bus);

fprintf('\n')
fprintf('Total active load demand (KW)                      :');
fprintf('%4.3f',Pload_KW);
fprintf('\n')
fprintf('Total reactive load demand (KVAR)                  :');
fprintf('%4.3f',Qload_KVAR);
fprintf('\n')
fprintf('Total Active Loss (KW)                             :');
fprintf('%4.3f ' ,TotalActiveLoss_KW);
fprintf('\n')
fprintf('Total Reactive Loss (KVAR)                         :');
fprintf('%4.3f',TotalReactiveLoss_KVR);
fprintf('\n')
fprintf('Total energy losses cost ($)                       :');
% fprintf('%4.3f',Total_energy_loss_cost);
```

```
% fprintf('\n')
% fprintf('The generated active power by DGs (KW)          :');
% fprintf('%4.5f ',Pgt);
% fprintf('\n')
% fprintf('The generated reactive power by DGs or CB(KVAR):');
% fprintf('%4.5f',Qgt);

fprintf('\n')
fprintf('========================================
==================================')
fprintf('\n')
 format long
 %%
format short

%%%%%%%%%%%%Main file %%%%%%%%%%%%%

clear all
clc
tic
SearchAgents_no=5; % Number of search agents

Function_name='F1'; % Name of the test function that can be from F1 to F23
(Table 1,2,3 in the paper)

Max_iteration=100; % Maximum numbef of iterations

% Load details of the selected benchmark function
[lb,ub,dim,fobj]=Get_Functions_details(Function_name);

[Best_score,Best_pos,cg_curve]=ALO(SearchAgents_no,Max_iteration,
lb,ub,dim,fobj);

semilogy(cg_curve,'Color','r')
title('Convergence curve')
xlabel('Iteration');
ylabel('Best score obtained so far');

axis tight
grid off
box on
legend('ALO')

display(['The best solution obtained by ALO is : ', num2str
(Best_pos)]);
display(['The best optimal value of the objective funciton found by ALO
is : ', num2str(Best_score)]);

  x=Best_pos
 FinalTestConstaints
```

```
Toc
%%%%%%%%%%%%%%%%%%%%%%%%%%%%%%%%%%%%%%%%%%%%%%%%%%%%
%%%%%%%%%%%%%%%%%%get function details%%%%%%%%%%%%%%%%%%

function [lb,ub,dim,fobj] = Get_Functions_details(F)

switch F
  case 'F1'
      fobj = @F1;
       lb = [0.02    0.02     1     1      1     1] ;
       ub = [1.18  1.18    227.09     227.09      170  170] ;
       dim= 6;
     end

end
% F1

function o = F1(x)
%%
Penetration_level = 1; % 50 of load level
%%
accuracy = 1e-6; maxiter = 4;
nbus = 118;
 busdata =[  1.0000          0      0      0      0
   2.0000    133.840  101.140     0      0
   3.0000     16.214   11.292     0      0
   4.0000     34.315   21.845     0      0
   5.0000     73.016   63.602     0      0
   6.0000    144.200   68.604     0      0
   7.0000    104.470   61.725     0      0
   8.0000     28.547   11.503     0      0
   9.0000     87.560   51.073     0      0
  10.0000    198.200  106.770     0      0
  11.0000    146.800   75.995     0      0
  12.0000     26.040   18.687     0      0
  13.0000     52.100   23.220     0      0
  14.0000    141.900  117.500     0      0
  15.0000     21.870   28.790     0      0
  16.0000     33.370   26.450     0      0
  17.0000     32.430   25.230     0      0
  18.0000     20.234   11.906     0      0
  19.0000   156.940    78.523     0      0
  20.0000   546.290   351.400     0      0
  21.0000   180.310   164.200     0      0
  22.0000    93.167    54.594     0      0
  23.0000    85.180    39.650     0      0
  24.0000   168.100    95.178     0      0
  25.0000   125.110   150.220     0      0
  26.0000    16.030    24.620     0      0
  27.0000    26.030    24.620     0      0
  28.0000   594.560   522.620     0      0
  29.0000   120.620    59.117     0      0
  30.0000   102.380    99.554     0      0
```

| | | | | |
|---|---|---|---|---|
| 31.0000 | 513.400 | 318.500 | 0 | 0 |
| 32.0000 | 475.250 | 456.140 | 0 | 0 |
| 33.0000 | 151.430 | 136.790 | 0 | 0 |
| 34.0000 | 205.380 | 83.302 | 0 | 0 |
| 35.0000 | 131.600 | 93.082 | 0 | 0 |
| 36.0000 | 448.400 | 369.790 | 0 | 0 |
| 37.0000 | 440.520 | 321.640 | 0 | 0 |
| 38.0000 | 112.540 | 55.134 | 0 | 0 |
| 39.0000 | 53.963 | 38.998 | 0 | 0 |
| 40.0000 | 393.050 | 342.600 | 0 | 0 |
| 41.0000 | 326.740 | 278.560 | 0 | 0 |
| 42.0000 | 536.260 | 240.240 | 0 | 0 |
| 43.0000 | 76.247 | 66.562 | 0 | 0 |
| 44.0000 | 53.520 | 39.760 | 0 | 0 |
| 45.0000 | 40.328 | 31.964 | 0 | 0 |
| 46.0000 | 39.653 | 20.758 | 0 | 0 |
| 47.0000 | 66.195 | 42.361 | 0 | 0 |
| 48.0000 | 73.904 | 51.653 | 0 | 0 |
| 49.0000 | 114.770 | 57.965 | 0 | 0 |
| 50.0000 | 918.370 | 1205.100 | 0 | 0 |
| 51.0000 | 210.300 | 146.660 | 0 | 0 |
| 52.0000 | 66.680 | 56.608 | 0 | 0 |
| 53.0000 | 42.207 | 40.184 | 0 | 0 |
| 54.0000 | 433.740 | 283.410 | 0 | 0 |
| 55.0000 | 62.100 | 26.860 | 0 | 0 |
| 56.0000 | 92.460 | 88.380 | 0 | 0 |
| 57.0000 | 85.188 | 55.436 | 0 | 0 |
| 58.0000 | 345.300 | 332.400 | 0 | 0 |
| 59.0000 | 22.500 | 16.830 | 0 | 0 |
| 60.0000 | 80.551 | 49.156 | 0 | 0 |
| 61.0000 | 95.860 | 90.758 | 0 | 0 |
| 62.0000 | 62.920 | 47.700 | 0 | 0 |
| 63.0000 | 478.800 | 463.740 | 0 | 0 |
| 64.0000 | 120.940 | 52.006 | 0 | 0 |
| 65.0000 | 139.110 | 100.340 | 0 | 0 |
| 66.0000 | 391.780 | 193.500 | 0 | 0 |
| 67.0000 | 27.741 | 26.713 | 0 | 0 |
| 68.0000 | 52.814 | 25.257 | 0 | 0 |
| 69.0000 | 66.890 | 38.713 | 0 | 0 |
| 70.0000 | 467.500 | 395.140 | 0 | 0 |
| 71.0000 | 594.850 | 239.740 | 0 | 0 |
| 72.0000 | 132.500 | 84.363 | 0 | 0 |
| 73.0000 | 52.699 | 22.482 | 0 | 0 |
| 74.0000 | 869.790 | 614.775 | 0 | 0 |
| 75.0000 | 31.349 | 29.817 | 0 | 0 |
| 76.0000 | 192.390 | 122.430 | 0 | 0 |
| 77.0000 | 65.750 | 45.370 | 0 | 0 |
| 78.0000 | 238.150 | 223.220 | 0 | 0 |
| 79.0000 | 294.550 | 162.470 | 0 | 0 |
| 80.0000 | 485.570 | 437.920 | 0 | 0 |
| 81.0000 | 243.530 | 183.030 | 0 | 0 |
| 82.0000 | 243.530 | 183.030 | 0 | 0 |
| 83.0000 | 134.250 | 119.290 | 0 | 0 |

```
 84.0000  22.710  27.960     0     0
 85.0000  49.513  26.515     0     0
 86.0000 383.780 257.160     0     0
 87.0000  49.640  20.600     0     0
 88.0000  22.473  11.806     0     0
 89.0000  62.930  42.960     0     0
 90.0000  30.670  34.930     0     0
 91.0000  62.530  66.790     0     0
 92.0000 114.570  81.748     0     0
 93.0000  81.292  66.526     0     0
 94.0000  31.733  15.960     0     0
 95.0000  33.320  60.480     0     0
 96.0000 531.280 224.850     0     0
 97.0000 507.030 367.420     0     0
 98.0000  26.390  11.700     0     0
 99.0000  45.990  30.392     0     0
100.0000 100.660  47.572     0     0
101.0000 456.480 350.300     0     0
102.0000 522.560 449.290     0     0
103.0000 408.430 168.460     0     0
104.0000 141.480 134.250     0     0
105.0000 104.430  66.024     0     0
106.0000  96.793  83.647     0     0
107.0000 493.920 419.340     0     0
108.0000 225.380 135.880     0     0
109.0000 509.210 387.210     0     0
110.0000 188.500 173.460     0     0
111.0000 918.030 898.550     0     0
112.0000 305.080 215.370     0     0
113.0000  54.380  40.970     0     0
114.0000 211.140 192.900     0     0
115.0000  67.009  53.336     0     0
116.0000 162.070  90.321     0     0
117.0000  48.785  29.156     0     0
118.0000  33.900  18.980     0     0];

linedata = [ 1.0000   1.0000  2.0000  0.0360  0.0130
  2.0000      2.0000  3.0000  0.0330  0.0119
  3.0000      2.0000  4.0000  0.0450  0.0162
  4.0000      4.0000  5.0000   0.0150  0.0540
  5.0000      5.0000  6.0000   0.0150  0.0540
  6.0000      6.0000  7.0000   0.0150  0.0125
  7.0000      7.0000  8.0000   0.0180  0.0140
  8.0000      8.0000  9.0000   0.0210  0.0630
  9.0000      2.0000 10.0000   0.1660  0.1344
 10.0000     10.0000 11.0000   0.1120  0.0789
 11.0000     11.0000 12.0000   0.1870  0.3130
 12.0000     12.0000 13.0000   0.1420  0.1512
 13.0000     13.0000 14.0000   0.1800  0.1180
 14.0000     14.0000 15.0000   0.1500  0.0450
 15.0000     15.0000 16.0000   0.1600  0.1800
 16.0000     16.0000 17.0000   0.1570  0.1710
 17.0000     11.0000 18.0000   0.2180  0.2850
```

| | | | | |
|---|---|---|---|---|
| 18.0000 | 18.0000 | 19.0000 | 0.1180 | 0.1850 |
| 19.0000 | 19.0000 | 20.0000 | 0.1600 | 0.1960 |
| 20.0000 | 20.0000 | 21.0000 | 0.1200 | 0.1890 |
| 21.0000 | 21.0000 | 22.0000 | 0.1200 | 0.0789 |
| 22.0000 | 22.0000 | 23.0000 | 1.4100 | 0.7230 |
| 23.0000 | 23.0000 | 24.0000 | 0.2930 | 0.1348 |
| 24.0000 | 24.0000 | 25.0000 | 0.1330 | 0.1040 |
| 25.0000 | 25.0000 | 26.0000 | 0.1780 | 0.1340 |
| 26.0000 | 26.0000 | 27.0000 | 0.1780 | 0.1340 |
| 27.0000 | 4.0000 | 28.0000 | 0.0150 | 0.0296 |
| 28.0000 | 28.0000 | 29.0000 | 0.0120 | 0.0276 |
| 29.0000 | 29.0000 | 30.0000 | 0.1200 | 0.2766 |
| 30.0000 | 30.0000 | 31.0000 | 0.2100 | 0.2430 |
| 31.0000 | 31.0000 | 32.0000 | 0.1200 | 0.0540 |
| 32.0000 | 32.0000 | 33.0000 | 0.1780 | 0.2340 |
| 33.0000 | 33.0000 | 34.0000 | 0.1780 | 0.2340 |
| 34.0000 | 34.0000 | 35.0000 | 0.1540 | 0.1620 |
| 35.0000 | 30.0000 | 36.0000 | 0.1870 | 0.2610 |
| 36.0000 | 36.0000 | 37.0000 | 0.1330 | 0.0990 |
| 37.0000 | 29.0000 | 38.0000 | 0.3300 | 0.1940 |
| 38.0000 | 38.0000 | 39.0000 | 0.3100 | 0.1940 |
| 39.0000 | 39.0000 | 40.0000 | 0.1300 | 0.1940 |
| 40.0000 | 40.0000 | 41.0000 | 0.2800 | 0.1500 |
| 41.0000 | 41.0000 | 42.0000 | 1.1800 | 0.8500 |
| 42.0000 | 42.0000 | 43.0000 | 0.4200 | 0.2436 |
| 43.0000 | 43.0000 | 44.0000 | 0.2700 | 0.0972 |
| 44.0000 | 44.0000 | 45.0000 | 0.3390 | 0.1221 |
| 45.0000 | 45.0000 | 46.0000 | 0.2700 | 0.1779 |
| 46.0000 | 35.0000 | 47.0000 | 0.2100 | 0.1383 |
| 47.0000 | 47.0000 | 48.0000 | 0.1200 | 0.0789 |
| 48.0000 | 48.0000 | 49.0000 | 0.1500 | 0.0987 |
| 49.0000 | 49.0000 | 50.0000 | 0.1500 | 0.0987 |
| 50.0000 | 50.0000 | 51.0000 | 0.2400 | 0.1581 |
| 51.0000 | 51.0000 | 52.0000 | 0.1200 | 0.0789 |
| 52.0000 | 52.0000 | 53.0000 | 0.4050 | 0.1458 |
| 53.0000 | 53.0000 | 54.0000 | 0.4050 | 0.1458 |
| 54.0000 | 29.0000 | 55.0000 | 0.3910 | 0.1410 |
| 55.0000 | 55.0000 | 56.0000 | 0.4060 | 0.1461 |
| 56.0000 | 56.0000 | 57.0000 | 0.4060 | 0.1461 |
| 57.0000 | 57.0000 | 58.0000 | 0.7060 | 0.5461 |
| 58.0000 | 58.0000 | 59.0000 | 0.3380 | 0.1218 |
| 59.0000 | 59.0000 | 60.0000 | 0.3380 | 0.1218 |
| 60.0000 | 60.0000 | 61.0000 | 0.2070 | 0.0747 |
| 61.0000 | 61.0000 | 62.0000 | 0.2470 | 0.8922 |
| 62.0000 | 1.0000 | 63.0000 | 0.0280 | 0.0418 |
| 63.0000 | 63.0000 | 64.0000 | 0.1170 | 0.2016 |
| 64.0000 | 64.0000 | 65.0000 | 0.2550 | 0.0918 |
| 65.0000 | 65.0000 | 66.0000 | 0.2100 | 0.0759 |
| 66.0000 | 66.0000 | 67.0000 | 0.3830 | 0.1380 |
| 67.0000 | 67.0000 | 68.0000 | 0.5040 | 0.3303 |
| 68.0000 | 68.0000 | 69.0000 | 0.4060 | 0.1461 |
| 69.0000 | 69.0000 | 70.0000 | 0.9620 | 0.7610 |
| 70.0000 | 70.0000 | 71.0000 | 0.1650 | 0.0600 |

```
 71.0000  71.0000  72.0000  0.3030  0.1092
 72.0000  72.0000  73.0000  0.3030  0.1092
 73.0000  73.0000  74.0000  0.2060  0.1440
 74.0000  74.0000  75.0000  0.2330  0.0840
 75.0000  75.0000  76.0000  0.5910  0.1773
 76.0000  76.0000  77.0000  0.1260  0.0453
 77.0000  64.0000  78.0000  0.5590  0.3687
 78.0000  78.0000  79.0000  0.1860  0.1227
 79.0000  79.0000  80.0000  0.1860  0.1227
 80.0000  80.0000  81.0000  0.2600  0.1390
 81.0000  81.0000  82.0000  0.1540  0.1480
 82.0000  82.0000  83.0000  0.2300  0.1280
 83.0000  83.0000  84.0000  0.2520  0.1060
 84.0000  84.0000  85.0000  0.1800  0.1480
 85.0000  79.0000  86.0000  0.1600  0.1820
 86.0000  86.0000  87.0000  0.2000  0.2300
 87.0000  87.0000  88.0000  0.1600  0.3930
 88.0000  65.0000  89.0000  0.6690  0.2412
 89.0000  89.0000  90.0000  0.2660  0.1227
 90.0000  90.0000  91.0000  0.2660  0.1227
 91.0000  91.0000  92.0000  0.2660  0.1227
 92.0000  92.0000  93.0000  0.2660  0.1227
 93.0000  93.0000  94.0000  0.2330  0.1150
 94.0000  94.0000  95.0000  0.4960  0.1380
 95.0000  91.0000  96.0000  0.1960  0.1800
 96.0000   96.0000  97.0000  0.1960  0.1800
 97.0000   97.0000  98.0000  0.1866  0.1220
 98.0000   98.0000  99.0000  0.0746  0.3180
 99.0000    1.0000 100.0000  0.0625  0.0265
100.0000 100.0000 101.0000  0.1501  0.2340
101.0000 101.0000 102.0000  0.1347  0.0888
102.0000 102.0000 103.0000  0.2307  0.1203
103.0000 103.0000 104.0000  0.4470  0.1608
104.0000 104.0000 105.0000  0.1632  0.0588
105.0000 105.0000 106.0000  0.3300  0.0990
106.0000 106.0000 107.0000  0.1560  0.0561
107.0000 107.0000 108.0000  0.3819  0.1374
108.0000 108.0000 109.0000  0.1626  0.0585
109.0000 109.0000 110.0000  0.3819  0.1374
110.0000 110.0000 111.0000  0.2445  0.0879
111.0000 110.0000 112.0000  0.2088  0.0753
112.0000 112.0000 113.0000  0.2301  0.0828
113.0000 100.0000 114.0000  0.6102  0.2196
114.0000 114.0000 115.0000  0.1866  0.1270
115.0000 115.0000 116.0000  0.3732  0.2460
116.0000 116.0000 117.0000  0.4050  0.3670
117.0000 117.0000 118.0000  0.4890  0.4380
];
    %%
    %hour  spring   summer  Atonomy  winter
Loading = [1.0000   0.7410   0.6920   0.6950   0.6870
   2.0000   0.7080   0.6550   0.6540   0.6510
   3.0000   0.6750   0.6190   0.6200   0.6180
```

```
  4.0000   0.6610   0.6070    0.6110   0.6100
  5.0000   0.6720   0.6180    0.6150   0.6250
  6.0000   0.7180   0.6490    0.6350   0.6720
  7.0000   0.8020   0.7330    0.7150   0.7620
  8.0000   0.8180   0.7900    0.7780   0.8070
  9.0000   0.8430   0.8080    0.8090   0.8260
 10.0000   0.8660   0.8230    0.8380   0.8420
 11.0000   0.8800   0.8320    0.8580   0.8520
 12.0000   0.8860   0.8360    0.8680   0.8580
 13.0000   0.8820   0.8370    0.8700   0.8580
 14.0000   0.8730   0.8240    0.8610   0.8460
 15.0000   0.8710   0.8180    0.8570   0.8420
 16.0000   0.8710   0.8130    0.8530   0.8430
 17.0000   0.8770   0.8080    0.8490   0.8480
 18.0000   0.8740   0.8010    0.8350   0.8490
 19.0000   0.8890   0.8100    0.8180   0.8710
 20.0000   0.9000   0.8320    0.8180   0.8940
 21.0000   0.9040   0.8530    0.8340   0.8720
 22.0000   0.8780   0.8220    0.8400   0.8220
 23.0000   0.8390   0.7790    0.7970   0.7760
 24.0000   0.7970   0.7410    0.7520   0.7360];

%%      winter      spring     summer     autumn
rradiance =     [0        0        0           0
       0          0          0        0
       0          0          0        0
       0          0          0        0
       0          0          0        0
       0          0          0        0
       0    13.8690    28.3120     0.0020
  0.0060   150.8170   205.7910    37.8490
 39.6730   366.5050   418.7880   215.6190
201.2860   553.9440   596.0590   397.8950
361.8990   717.6510   757.6500   545.7280
474.4810   825.2850   860.9530   636.3370
516.1320   878.4380   881.4250   663.3410
506.2450   852.4560   837.0310   626.1620
425.1740   766.8170   770.8930   525.0770
293.1390   622.7320   607.0090   368.3480
108.5580   430.6630   468.4420   132.3930
  0.7270   189.0940   300.6030    12.5290
       0    32.5410   113.1220     0.4930
       0     0.0020     0.0730          0
       0          0          0          0
       0          0          0          0
       0          0          0          0
       0          0          0          0];
%%
Gstd=1000; % standard of solar
Rc=120;
Gs=rradiance ;
Gs_winter = Gs(:,1) ;
Gs_Spring = Gs(:,2) ;
```

```
Gs_Summer = Gs(:,3) ;
Gs_Atonomy = Gs(:,4);

%%
LL1 = round(100*x(1));
LL2 = round(100*x(2));
%
L1 = LL1;  % Location of frist PV1
L2 = LL2;  % Location of frist PV1
L3 = L1;   % Location of frist D-STATCOM
L4 = L2;   % Location of frist D-STATCOM

% L1 = (L1,2);    % Location of frist pv
% L2 = (LL2,2);   % Location of frist D-STATCOM

PSR1 = round((100*x(3)));
PSR2 = round((100*x(4)));

Q_D-STATCOM1 = round((100*x(5)));
Q_D-STATCOM2 = round((100*x(6)));

busdata(L3,5) = Q_D-STATCOM1;
busdata(L4,5) = Q_D-STATCOM2;
 %%
% %% DSTAT^COM cost
Q_STATCOM = Q_D-STATCOM1;
B = 0.1;
K_Statcom = 50;
N_Statcom = 30;
AA = (1+B)^N_Statcom;
Statcom_Cost = Q_STATCOM*K_Statcom *((AA*B)/(AA-1));
%%
orginal_busdata = busdata;
winter_loading = Loading(:,2);
spring_loading = Loading(:,3);
summer_loading = Loading(:,4);
Atonomy_loading = Loading(:,5);
Total_PL = sum (orginal_busdata(:,2));
Total_QL = sum (orginal_busdata(:,3));
%%

%%
%%%%%% spring_loading %%%%%
% disp('%%%%%% spring_loading %%%%%')
for hh = 1:length(spring_loading)

 busdata(:,2) =  spring_loading(hh)*orginal_busdata(:,2);
 busdata(:,3) =  spring_loading(hh)*orginal_busdata(:,3);

 if  Gs_Spring(hh)  < Rc
   PS_Spring1(hh)=PSR1*(((Gs_Spring(hh))^2)/(Gstd*Rc));
   PS_Spring2(hh)=PSR1*(((Gs_Spring(hh))^2)/(Gstd*Rc));
```

```
elseif  Gs_Spring(hh) > Rc || Gs_Spring(hh) == Rc
   PS_Spring1(hh) = PSR1 * (Gs_Spring(hh)/Gstd);
   PS_Spring2(hh) = PSR2 * (Gs_Spring(hh)/Gstd);
elseif Gs_Spring(hh) > Gstd
 PS_Spring1(hh)=PSR1  ;
 PS_Spring2(hh)=PSR2 ;

 end

 busdata(L1,4) = PS_Spring1(hh);
 busdata(L2,4) = PS_Spring2(hh);

PerUnit ;
backward_forward_sweep ;
Calculation1 ;
Plloss_spring(hh) = TotalActiveLoss_KW;
VDD_spring(hh) = VDD;
VSI_spring(hh)  = Sum_VSI;
Real_Power_Supplied_From_Substation_KW_spring(hh) =
Real_Power_Supplied_From_Substation_KW;

 %__________________ Constraints ____________________

for i = 1: length(V_mag_no1)
if V_mag_no1(i) > 1.05 || V_mag_no1(i) < 0.90
  k1(i) = 1000;
else
  k1(i) = 0;
end
end
k1 = sum(k1);
k11(hh) = sum(k1);

end
PS_Spring_out1 = sum (PS_Spring1)*91.25;
PS_Spring_out2 = sum (PS_Spring2)*91.25;

PS_Spring_out  = PS_Spring_out1+PS_Spring_out2;

Plloss_spring;
Sum_Plloss_spring  = sum(Plloss_spring) ;
E_spring = Sum_Plloss_spring *91.25 ;
Sum_VD_spring   = sum(VDD_spring) ;
Sum_VSI_spring  = sum(VSI_spring) ;
E_Station_spring =  sum
(Real_Power_Supplied_From_Substation_KW_spring )*91.25;

%%
%%%%%% summer_loading %%%%%
% disp('%%%%%% summer_loading %%%%%')
for hh = 1:length(summer_loading)
 hh ;
```

```
busdata(:,2) = summer_loading(hh)*orginal_busdata(:,2);
busdata(:,3) = summer_loading(hh)*orginal_busdata(:,3);

 if  Gs_Summer(hh)  < Rc
  PS_Summer1(hh)=PSR1*(((Gs_Summer(hh))^2)/(Gstd*Rc));
  PS_Summer2(hh)=PSR2*(((Gs_Summer(hh))^2)/(Gstd*Rc));

elseif  Gs_Summer(hh)>=Rc  && Gs_Summer(hh)<= Gstd
 PS_Summer1(hh)=PSR1 * (Gs_Summer(hh)/Gstd);
 PS_Summer2(hh)=PSR2 * (Gs_Summer(hh)/Gstd);

 elseif Gs_Summer(hh)> Gstd
 PS_Summer1(hh)=PSR1   ;
 PS_Summer2(hh)=PSR2  ;

 end

 busdata(L1,4) = PS_Summer1(hh);
 busdata(L2,4) = PS_Summer2(hh);

PerUnit
backward_forward_sweep
Calculation1
Plloss_summer(hh) = TotalActiveLoss_KW;
VDD_summer(hh) = VDD;
VSI_summer(hh)  = Sum_VSI;
Real_Power_Supplied_From_Substation_KW_summer(hh)  =
Real_Power_Supplied_From_Substation_KW;

 %__________________ Constraints ____________________

for i = 1: length(V_mag_no1)
if V_mag_no1(i)> 1.05 || V_mag_no1(i)< 0.90
  k2(i) = 1000;
else
  k2(i) = 0;
end
end
k2 = sum(k2);
k22(hh) = sum(k2);
end
PS_Summer_out1 = sum (PS_Summer1)*91.25;
PS_Summer_out2 = sum (PS_Summer2)*91.25;
PS_Summer_out = PS_Summer_out1+PS_Summer_out2;

Plloss_summer;
Sum_Plloss_summer = sum(Plloss_summer);
E_summer = Sum_Plloss_summer *91.25 ;
Sum_VD_summer  = sum(VDD_summer);
Sum_VSI_summer = sum(VSI_summer);
E_Station_summer = sum
(Real_Power_Supplied_From_Substation_KW_summer)*91.25;
```

```
%%%%%% Atonomy_loading %%%%%
% disp('%%%%%% Atonomy_loading %%%%%')
for hh = 1:length(Atonomy_loading)
 hh;
 busdata(:,2) = Atonomy_loading(hh)*orginal_busdata(:,2);
 busdata(:,3) = Atonomy_loading(hh)*orginal_busdata(:,3);

if Gs_Atonomy(hh) < Rc
   PS_Atonomy1(hh)= PSR1*(((Gs_Atonomy(hh))^2)/(Gstd*Rc));
   PS_Atonomy2(hh)= PSR2*(((Gs_Atonomy(hh))^2)/(Gstd*Rc));

 elseif Gs_Atonomy(hh)>=Rc && Gs_Atonomy(hh)<= Gstd
   PS_Atonomy1(hh)= PSR1 * (Gs_Atonomy(hh)/Gstd);
   PS_Atonomy2(hh)= PSR2 * (Gs_Atonomy(hh)/Gstd);

  elseif Gs_Atonomy(hh)> Gstd
  PS_Atonomy1(hh)=PSR1;
  PS_Atonomy2(hh)=PSR2;

 end
  busdata(L1,4) = PS_Atonomy1(hh);
  busdata(L2,4) = PS_Atonomy2(hh);

PerUnit
backward_forward_sweep
Calculation1
Plloss_Atonomy(hh) = TotalActiveLoss_KW;
VDD_Atonomy(hh) = VDD;
VSI_Atonomy(hh) = Sum_VSI;
Real_Power_Supplied_From_Substation_KW_Atonomy(hh) =
Real_Power_Supplied_From_Substation_KW;

 %___________________ Constraints ____________________

for i = 1: length(V_mag_no1)
if V_mag_no1(i)> 1.05 || V_mag_no1(i)< 0.90
  k3(i) = 1000;
else
  k3(i) = 0;
end
end
k3 = sum(k3);
k33(hh) = sum(k3);
end
PS_Atonomy_out1 = sum (PS_Atonomy1)*91.25;
PS_Atonomy_out2 = sum (PS_Atonomy2)*91.25;
PS_Atonomy_out  = PS_Atonomy_out1+PS_Atonomy_out2;

Plloss_Atonomy;
Sum_Plloss_Atonomy = sum(Plloss_Atonomy );

%%
```

```
E_Atonomy = Sum_Plloss_Atonomy *91.25 ;
Sum_VD_Atonomy  = sum(VDD_Atonomy);
Sum_VSI_Atonomy = sum(VSI_Atonomy);
E_Station_Atonomy = sum
(Real_Power_Supplied_From_Substation_KW_Atonomy)*91.25;

%%
%%%%%% winter_loading %%%%%
% disp('%%%%%% winter_loading %%%%%')
for hh = 1:length(winter_loading)
 hh ;
 busdata(:,2) = winter_loading(hh)*orginal_busdata(:,2);
 busdata(:,3) = winter_loading(hh)*orginal_busdata(:,3);
if  Gs_winter(hh)  < Rc
   PS_Winter1(hh) = PSR1*(((Gs_winter(hh))^2)/(Gstd*Rc));
   PS_Winter2(hh) = PSR2*(((Gs_winter(hh))^2)/(Gstd*Rc));

 elseif  Gs_winter(hh) >=Rc  && Gs_winter(hh) <= Gstd
   PS_Winter1(hh) = PSR1 * (Gs_winter(hh)/Gstd);
   PS_Winter2(hh) = PSR2 * (Gs_winter(hh)/Gstd);

  elseif Gs_winter(hh) > Gstd
  PS_Winter1(hh) =PSR1;
  PS_Winter2(hh) =PSR2;

 end

busdata(L1,4) = PS_Winter1(hh);
busdata(L2,4) = PS_Winter2(hh);

PerUnit
backward_forward_sweep
Calculation1
Plloss_winter(hh) = TotalActiveLoss_KW;
VDD_winter(hh) = VDD;
VSI_winter(hh)  = Sum_VSI;
Real_Power_Supplied_From_Substation_KW_winter(hh) =
Real_Power_Supplied_From_Substation_KW;

 %__________________ Constraints ____________________

for i = 1: length(V_mag_no1)
if V_mag_no1(i) > 1.05 || V_mag_no1(i) < 0.90
   k4(i) = 1000;
else
   k4(i) = 0;
end
end
k4 = sum(k4);
k44(hh) = sum(k4);
k44(hh);
end
```

```
PS_Winter_out1 = sum (PS_Winter1)*91.25;
PS_Winter_out2 = sum (PS_Winter2)*91.25;
PS_Winter_out  = PS_Winter_out1+PS_Winter_out2;

Plloss_winter;
Sum_Plloss_winter  = sum(Plloss_winter) ;
E_winter = Sum_Plloss_winter *91.25 ;
Sum_VD_winter   = sum(VDD_winter);
Sum_VSI_winter  = sum(VSI_winter);
E_Station_winter =  sum
(Real_Power_Supplied_From_Substation_KW_winter)*91.25;

%%
%%%%%%%%%%%%%%%%%%%%%%%%
 Power_PV =PSR1 + PSR2;
%  % Cost of PV
B = 0.1;
CSDG = 770;
N_PV = 20;
AA = (1+B)^N_PV;
Capital_cost = Power_PV*CSDG * ((AA*B)/(AA-1));

Total_KWh_PV = PS_Winter_out + PS_Atonomy_out + PS_Summer_out +
PS_Spring_out;
a = Capital_cost ;
O_M_cost  = 0.01;
Fuel_cost = 0;
b =  O_M_cost +Fuel_cost;
Cost_PV = a+ b*Total_KWh_PV;

%%    result
% Penteration level constraints
Pload = Total_PL*Penetration_level;

if Power_PV > 22709
  m5 = 1000000;
else
  m5 = 0;
end
if Q_STATCOM > 17000
  m6 = 1000000;
else
  m6 = 0;
end
%%    result
%%
Power_PV =PSR1+PSR2 ;
%  % Cost of PV
B = 0.1;
CSDG = 770;
N_PV = 20;
```

```
AA = (1+B)^N_PV;
Capital_cost = Power_PV*CSDG * ((AA*B)/(AA-1));

Total_KWh_PV = PS_Winter_out + PS_Atonomy_out + PS_Summer_out +
PS_Spring_out;
a = Capital_cost ;
O_M_cost  = 0.01;
Fuel_cost = 0;
b = O_M_cost +Fuel_cost;
Cost_PV = a+ b*Total_KWh_PV;
%%    result
% Penteration level constraints
Pload = Total_PL;
% *Penetration_level;

if Power_PV > 22709
  m5 = 1000000;
else
  m5 = 0;
end
if Q_STATCOM > 17000
  m6 = 1000000;
else
  m6 = 0;
end
%
%%
format long

Total_E_from_substation =  E_Station_spring + E_Station_summer...
                        + E_Station_Atonomy + E_Station_winter ;
Total_Enery_loss = E_spring + E_summer + E_Atonomy + E_winter;

Cost_Total_Enery_loss = 0.06*Total_Enery_loss ;
Ke =  0.096;
Purchasing_power_from_Substation = Total_E_from_substation * Ke;

Total_VD = 91.25*(Sum_VD_spring + Sum_VD_summer +Sum_VD_Atonomy +
Sum_VD_winter);

Total_VSI = 91.25*(Sum_VSI_spring  + Sum_VSI_summer +Sum_VSI_Atonomy
+Sum_VSI_winter);
Ke =  0.096;
m1= sum(k11) ;
m2=sum(k22) ;
m3=sum(k33) ;
m4=sum(k44);
m5;
m6;
All_over_cost = Cost_Total_Enery_loss +
Purchasing_power_from_Substation +Statcom_Cost +Cost_PV;
```

```
Statcom_Cost;
Cost_PV;

%o = All_over_cost+ 1000*m1 + 1000*m2 +1000*m3 +1000*m4 +100000*m5+ m6;
optimal_location_hybrid1 = L1;
optimal_location_hybrid2 = L2;
optimal_location_D-STATCOM_Hybrid1 = L1;
optimal_location_D-STATCOM_Hybrid2 = L2;
Optimal_size_KW_PV1 = PSR1;
Optimal_size_KW_PV2 = PSR2 ;
optimal_size_D-STATCOM_hybrid2 = Q_D-STATCOM1;
optimal_size_D-STATCOM_hybrid2 = Q_D-STATCOM2;

%%

Net_Saving = 1.205051854113218e+07-All_over_cost;
O1 = All_over_cost/1.205051854113218e+07;
O2 = Total_VD/ 3.5605e+04;
O3 = 1/Total_VSI;

O3 = 1/Total_VSI;% O4= Total_Enery_loss/4.935812577976250e+05;
w1=0.5;
w2=0.25;
w3=0.25;
o = w1*O1+w2*O2+w3*O3+ 1000*m1 +1000*m2 +1000*m3 +10000*m4+1000*m5
+m6;

end
%%%%%%%%%%%%%%%%%%%%%%%%%%%%%%%%%%%%%%%%%%%%%%%%
%%%%%%%ALO ALGORTHEM %%%%%%%%%%%%%%%%%%%%%%%%
%_____________________________________________________________________%
% Ant Lion Optimizer (ALO) source codes demo version 1.0              %
%                                                                      %
% Developed in MATLAB R2011b(7.13)                                    %
%                                                                      %
% Author and programmer: Seyedali Mirjalili                           %
%                                                                      %
%      e-Mail: ali.mirjalili@gmail.com                                %
%              seyedali.mirjalili@griffithuni.edu.au                  %
%                                                                      %
%     Homepage: http://www.alimirjalili.com                            %
%                                                                      %
% Main paper:                                                          %
%                                                                      %
% S. Mirjalili, The Ant Lion Optimizer                                 %
% Advances in Engineering Software , in press,2015                    %
% DOI: http://dx.doi.org/10.1016/j.advengsoft.2015.01.010              %
%                                                                      %
%_____________________________________________________________________%

% You can simply define your cost in a seperate file and load its handle to
fobj
```

```
% The initial parameters that you need are:
%__________________________________________
% fobj = @YourCostFunction
% dim = number of your variables
% Max_iteration = maximum number of generations
% SearchAgents_no = number of search agents
% lb=[lb1,lb2,...,lbn] where lbn is the lower bound of variable n
% ub=[ub1,ub2,...,ubn] where ubn is the upper bound of variable n
% If all the variables have equal lower bound, you can just
% define lb and ub as two single number numbers

% To run ALO: [Best_score,Best_pos,cg_curve]=ALO(SearchAgents_no,
Max_iteration,lb,ub,dim,fobj)

function [Elite_antlion_fitness,Elite_antlion_position,
Convergence_curve]=ALO(N,Max_iter,lb,ub,dim,fobj)

% Initialize the positions of antlions and ants
antlion_position=initialization(N,dim,ub,lb);
ant_position=initialization(N,dim,ub,lb);

% Initialize variables to save the position of elite, sorted antlions,
% convergence curve, antlions fitness, and ants fitness
Sorted_antlions=zeros(N,dim);
Elite_antlion_position=zeros(1,dim);
Elite_antlion_fitness=inf;
Convergence_curve=zeros(1,Max_iter);
antlions_fitness=zeros(1,N);
ants_fitness=zeros(1,N);

% Calculate the fitness of initial antlions and sort them
for i=1:size(antlion_position,1)
  antlions_fitness(1,i)=fobj(antlion_position(i,:));
end

[sorted_antlion_fitness,sorted_indexes]=sort(antlions_fitness);

for newindex=1:N
    Sorted_antlions(newindex,:)=antlion_position(sorted_indexes
(newindex),:);
end

Elite_antlion_position=Sorted_antlions(1,:);
Elite_antlion_fitness=sorted_antlion_fitness(1);

% Main loop start from the second iteration since the first iteration
% was dedicated to calculating the fitness of antlions
Current_iter=2;
while Current_iter<Max_iter+1
```

```
  % This for loop simulate random walks
  for i=1:size(ant_position,1)
    % Select ant lions based on their fitness (the better anlion the higher
chance of catching ant)
      Rolette_index=RouletteWheelSelection(1./
sorted_antlion_fitness);
      if Rolette_index==-1
         Rolette_index=1;
      end

      % RA is the random walk around the selected antlion by rolette wheel
      RA=Random_walk_around_antlion(dim,Max_iter,lb,ub,
Sorted_antlions(Rolette_index,:),Current_iter);

      % RA is the random walk around the elite (best antlion so far)
      [RE]=Random_walk_around_antlion(dim,Max_iter,lb,ub,
Elite_antlion_position(1,:),Current_iter);

    ant_position(i,:)= (RA(Current_iter,:)+RE(Current_iter,:))/2; %
Equation (2.13) in the paper
  end

  for i=1:size(ant_position,1)

      % Boundar checking (bring back the antlions of ants inside search
      % space if they go beyoud the boundaries
      Flag4ub=ant_position(i,:)>ub;
      Flag4lb=ant_position(i,:)<lb;
      ant_position(i,:)=(ant_position(i,:).*(~(Flag4ub+Flag4lb)))
+ub.*Flag4ub+lb.*Flag4lb;

      ants_fitness(1,i)=fobj(ant_position(i,:));

  end

  % Update antlion positions and fitnesses based of the ants (if an ant
  % becomes fitter than an antlion we assume it was cought by the antlion
  % and the antlion update goes to its position to build the trap)
  double_population=[Sorted_antlions;ant_position];
  double_fitness=[sorted_antlion_fitness ants_fitness];

  [double_fitness_sorted I]=sort(double_fitness);
  double_sorted_population=double_population(I,:);

  antlions_fitness=double_fitness_sorted(1:N);
  Sorted_antlions=double_sorted_population(1:N,:);

  % Update the position of elite if any antlinons becomes fitter than it
  if antlions_fitness(1)<Elite_antlion_fitness
     Elite_antlion_position=Sorted_antlions(1,:);
     Elite_antlion_fitness=antlions_fitness(1);
  end
```

```
  % Keep the elite in the population
  Sorted_antlions(1,:)=Elite_antlion_position;
  antlions_fitness(1)=Elite_antlion_fitness;

  % Update the convergence curve
  Convergence_curve(Current_iter)=Elite_antlion_fitness;

  % Display the iteration and best optimum obtained so far
  if mod(Current_iter,50)==0
    display(['At iteration ', num2str(Current_iter), ' the elite fitness
is ', num2str(Elite_antlion_fitness)]);
  end

  Current_iter=Current_iter+1;
End
%%%%%%%%%%%%%%%%%%%%%%%%%%%%%%%%%%%%%%%%%%%%%%%%%
CULCLUATION
% calculation
fprintf
('===============================================')
fprintf('\n')
Vbus
fprintf('\n')
IB

V_mag = abs(Vbus);
V_angle = angle(Vbus)*180/pi;
I_mag = abs(IB);
I_angle = angle(IB)*180/pi;
I_mag_square = I_mag.^2;

Linenumber = LD(:,1);
sending = LD(:,2);
receving = LD(:,3);

Busnumber = BD(:,1);

Ploss = I_mag_square.*R;
Qloss = I_mag_square.*X;
 for i = 1:length(busdata(:,1))
   VD(i) = (abs(V_mag(i)-1));
   VD2(i) = (abs(V_mag(i)-1))^2;
 end
VDD = sum(VD)
VDD2 = sum(VD2)
format long
V_mag
format short

 [V_mag_no1 I]=sort(V_mag);
V_mag_no1;
MinVoltage = V_mag_no1(1)
```

```
MaxVoltage = V_mag_no1((nbus-1))
Min_bus = I(1);
Max_bus = I(nbus-1);

Pload = BD (:,2)*1000* MVAb;
Qload = BD (:,3)*1000*MVAb;

%% -------------------------------------------
%Calculation voltage stability index

sending = LD(:,2);
receving = LD(:,3);
VSI(1) = 0;
Sef(1) = 0;
for i = 1: length(receving)
P_effective(receving(i)) = real(IB(i)*V_mag(receving(i)));
Q_effective(receving(i)) = -imag(IB(i)*V_mag(receving(i)));
end

for i = 1: length(receving)
VSI(receving(i)) = (V_mag(sending(i)))^4-4*(P_effective(receving
(i))*X(i)-Q_effective(receving(i))*R(i))^2 ...
-4*(P_effective(receving(i))*R(i)+Q_effective(receving(i))*X(i))*
(V_mag(sending(i)))^2 ;

end
VSI(1) = [];
 bus_stability_index=[receving VSI']
VSI_min = min(VSI)
Sum_VSI = sum(VSI)
%%------------------------------------------------
%Loss sensitivity factors (Reactive)
LSF(1) = 0;
for i = 1: length(receving)
LSF(receving(i)) = 2*Q_effective(receving(i))*R(i)/(V_mag(receving
(i)))^2;
end
[a candidate_location] = sort(LSF);

Q_Loss_sensetivity_index = [candidate_location' a']
%%--------------------------------
-------------------------------
%normalize
Vnormal = V_mag/0.95;

for i= 1:length(candidate_location)
 Voltage = Vnormal(candidate_location) ;

end
 fprintf('\n')
fprintf('========================================')
fprintf('\n')
```

```
fprintf('  Bus no.   LSI    Vn(|V/0.95|)     ')
fprintf('\n')
fprintf('=============================================')
fprintf('\n')

 LSSf = [candidate_location' a' Voltage ]

 %%
 fprintf('\n')
 %Loss sensitivity factors (active)
PLSF(1) = 0;
for i = 1: length(receving)
PLSF(receving(i)) = 2*P_effective(receving(i))*R(i)/(V_mag
(receving(i)))^2;
end

[aa candidate_locationn] = sort(PLSF);

Power_Loss_sensetivity_index = [candidate_locationn' aa']

 %%
 fprintf('\n')
 %Loss sensitivity factors (voltage)
LSI1(1) = 0;
for i = 1: length(receving)
LSI1(receving(i)) =  -2*R(i)*(((P_effective(receving(i)))
^2+(Q_effective(receving(i)))^2)/(V_mag(receving(i)))^3);
end

[av candidate_locationv] = sort(LSI1);
 format long
Voltage_Loss_sensetivity_index = [candidate_locationv' av']
format short

%%
fprintf('\n')
fprintf('============================================')
fprintf('\n')
fprintf('  bus no.  mag_V   angle_V     Pg_KW   Qg_KVAR ')
fprintf('\n')
fprintf
('===============================================')
fprintf('\n')
BusOut  = [Busnumber V_mag  V_angle  Pg_KW_bus     Qg_KVAR_bus ]
%Pload_KW_bus   Qload_KVAR_bus   Sload_KVA_bus  PowerFactor

PlossKW = Ploss*1000*MVAb;
QlossKVAR = Qloss*1000*MVAb;
fprintf('\n')
fprintf
('================================================')
fprintf('\n')
```

```
fprintf('  Line no.  from    to    mag_I(p.u) Ploss(KW) Qloss(KVAR)    ')
fprintf('\n')
fprintf
('==================================================')
fprintf('\n')

LineFlow = [Linenumber sending receving I_mag PlossKW QlossKVAR ]
TotalActiveLoss_KW = sum(PlossKW);
TotalReactiveLoss_KVR = sum(QlossKVAR);
Real_Power_Supplied_From_Substation_KW = TotalActiveLoss_KW +
Pload_KW - Pgt;
Reactive_Power_Supplied_From_Substation_KVAR =
TotalReactiveLoss_KVR + Qload_KVAR -Qgt;
Pt = Real_Power_Supplied_From_Substation_KW;
Qt = Real_Power_Supplied_From_Substation_KW;
%% Calculate the energy cost
Total_energy_loss_cost = TotalActiveLoss_KW*T*0.06;

%% Calculate the capacitor cost
Kc = 3; % purchase cost per KVAR
Kb = 1000; %installation cost
kd = 300; % operating cost
Capacitors_purchasing_cost = Qgt*Kc;
kk = 0;
% detrmine number of selecting bus for capacitor
for i = 1:length(Qg_KVAR_bus)
 if Qg_KVAR_bus(i) > 0
    kk = kk +1;
 end
end
Capacitors_installation_cost = kk* Kb  ;
Capacitors_operating_cost  = kk* kd;
Total_Capacitors_cost = Capacitors_purchasing_cost +
Capacitors_installation_cost +Capacitors_operating_cost;
Total_Capacitors_cost_without_operating_cost =
Capacitors_purchasing_cost + Capacitors_installation_cost;
Total_cost_cost_without_operating_cost =
Total_Capacitors_cost_without_operating_cost +
Total_energy_loss_cost;
Total_cost = Total_Capacitors_cost + Total_energy_loss_cost;

%%
fprintf('\n')
fprintf
('==================================================')
fprintf('\n')
fprintf('Minimum voltage (p.u)                        :');
fprintf('%3.5f',MinVoltage);
fprintf('  @ bus ');
fprintf('%4.f',Min_bus);

fprintf('\n')
fprintf('Maximum voltage (p.u) excluding the slack bus  :');
```

```
fprintf('%4.5f',MaxVoltage);
fprintf(' @bus ');
fprintf('%4.f',Max_bus);

fprintf('\n')
fprintf('Total active load demand (KW)                :');
fprintf('%4.3f',Pload_KW);
fprintf('\n')
fprintf('Total reactive load demand (KVAR)            :');
fprintf('%4.3f',Qload_KVAR);
fprintf('\n')
fprintf('Total Active Loss (KW)                       :');
fprintf('%4.3f ' ,TotalActiveLoss_KW);
fprintf('\n')
fprintf('Total Reactive Loss (KVAR)                   :');
fprintf('%4.3f',TotalReactiveLoss_KVR);
fprintf('\n')
fprintf('Total energy losses cost ($)                 :');
fprintf('%4.3f',Total_energy_loss_cost);
fprintf('\n')
fprintf('The generated active power by DGs (KW)       :');
fprintf('%4.5f ',Pgt);
fprintf('\n')
fprintf('The generated reactive power by DGs or CB(KVAR):');
fprintf('%4.5f',Qgt);
fprintf('\n')
 fprintf('Capacitors purchasing cost(KVAR)            :');
fprintf('%4.5f',Capacitors_purchasing_cost);
fprintf('\n')
fprintf('Capacitors installation cost                 :');
fprintf('%4.5f',Capacitors_installation_cost);
fprintf('\n')
fprintf('Capacitors operating cost                    :');
fprintf('%4.5f',Capacitors_operating_cost);
fprintf('\n')
fprintf('Total Capacitors cost                        :');
fprintf('%4.5f',Total_Capacitors_cost);
fprintf('\n')
fprintf('Total Capacitors cost without operating cost  :');
fprintf('%4.5f',Total_Capacitors_cost_without_operating_cost);
fprintf('\n')
fprintf('Total cost of energy and cpacitors without operating cost :');
fprintf('%4.5f',Total_cost_cost_without_operating_cost);
fprintf('\n')
fprintf('Total cost (energy cost + capacitors cost)               :');
fprintf('%4.5f',Total_cost);
fprintf('\n')
fprintf('=========================================')
fprintf('\n')

%%____________ power flow in transmission line________________
t=1:1:length(V_mag);
bar(t,LSF);
```

```
grid on
xlabel('Bus Number')
ylabel('LSF')
%%%%%%%%%%%%%%%%%%%%%%%%%%%%%%%%%%%%%%%%%%%%%%%%%%
%%%%%%%%%%%%%%backward forward sweep %%%%%%%%%%%

F = LD(:,2:3);
M=max(LD(:,2:3));
N=max(M);
f=[1:N]';
for i=1:N
  g=find(F(:,:)==i);
  h(i)=length(g);
end
k(:,1)=f;
k(:,2)=h';
 cent=1;%input('central bus ')
% this section of the code is to adjust line data to the standard
NLD = zeros(N,size(LD,2));
c=find(LD(:,2:3)==cent);
NLD=LD(c,:) ;
LD(c,:)=[];
t=find(k(:,1)==cent);
k(t,2)=k(t,2)-size(c,1);
j=size(c,1);
i=1 ;
while sum(k( ))>0
  c=[];
  b=[];
  t=[];
    if i >=85 && nbus ==85;
      i= 84;
    end
%     if nbus == 15 && i >=15
%
%       i= 14;
%     end
%
   [c e]=find(LD(:,2:3)==NLD(i,3));
  if size(c,2)~=0
     b=LD(c,:);
     LD(c,:)=[];
     t=find(k(:,1)==NLD(i,3));
     k(t,2)=k(t,2)-(size(c,1)+1);
     d=find(b(:,3)==NLD(i,3));
     b(d,2:3)=[b(d,3),b(d,2)];
     NLD(j+1:j+size(c,1),:)=b;
     j=j+size(c,1);
  end
  i=i+1;
end
LD=sortrows(NLD,3);
% end the data is represented in standard format
```

```
%code for bus-injection to branch-current matrix
bibc=zeros(size(LD,1),size(LD,1));
for i=1:size(LD,1)

  if LD(i,2)==1
    bibc(LD(i,3)-1,LD(i,3)-1)=1;
  else
    bibc(:,LD(i,3)-1)=bibc(:,LD(i,2)-1);
    bibc(LD(i,3)-1,LD(i,3)-1)=1;
  end
end
bibc;

S=complex(BD(:,2)-BD(:,4),BD(:,3)-BD(:,5)); % complex power
Vo=ones(size(LD,1),1); % initial bus votage% 10 change to specific data
value
S(1)=[];
VB=Vo;
iteration = maxiter;          %input('number of iteration : ');
for i=1:iteration
   %backward sweep
  I=conj(S./VB);% injected current
  Z=complex(LD(:,4),LD(:,5));%branch impedance
  ZD=diag(Z); %makeing it diagonal
  IB=bibc*I; %branch current
  %forward sweep
  TRX=bibc'*ZD*bibc;
  VB=Vo-TRX*I;
end

Vbus=[1;VB];
```

# References

1. L.A. Gallego, J.F. Franco, L.G. Cordero, A fast-specialized point estimate method for the probabilistic optimal power flow in distribution systems with renewable distributed generation. Int. J. Electr. Power Energy Syst. **131**, 107049 (2021)
2. S. Barakat, H. Ibrahim, A.A. Elbaset, Multi-objective optimization of grid-connected PV-wind hybrid system considering reliability, cost, and environmental aspects. Sustain. Cities Soc. **60**, 102178 (2020)
3. G. Magdy, G. Shabib, A.A. Elbaset, Y. Mitani, A comprehensive digital protection scheme for low-inertia microgrids considering high penetration of renewables, in *Renewable Power Systems Dynamic Security*, (Springer, 2020), pp. 39–57
4. F. Shahnia, A. Arefi, G. Ledwich, *Electric Distribution Network Planning* (Springer, 2018)
5. A. Oskoui, B. Mathew, J. Hasler, M. Oliveira, T. Larsson, A. Petersson, E. John, Holly STATCOM-FACTS to replace critical generation, in *Operational Experience, IEEE Power and Energy Society Transmission and Distribution Conference and Exposition*, (2005)
6. S. Jazebi, S.H. Hosseinian, B. Vahidi, DSTATCOM allocation in distribution networks considering reconfiguration using differential evolution algorithm. Energy Convers. Manag. **52**, 2777–2783 (2011)
7. S. Kamel, F. Jurado, D. Vera, A simple implementation of power mismatch STATCOM model into current injection Newton–Raphson power-flow method. Electr. Eng. **96**, 135–144 (2014)
8. G. Pepermans, J. Driesen, D. Haeseldonckx, R. Belmans, W. D'haeseleer, Distributed generation: Definition, benefits and issues. Energy Policy **33**, 787–798 (2005)
9. E.S. Oda, A.M. Abd El Hamed, A. Ali, A.A. Elbaset, M. Abd El Sattar, M. Ebeed, Stochastic optimal planning of distribution system considering integrated photovoltaic-based DG and DSTATCOM under uncertainties of loads and solar irradiance. IEEE Access **9**, 26541–26555 (2021)
10. R.O. Bawazir, N.S. Cetin, Comprehensive overview of optimizing PV-DG allocation in power system and solar energy resource potential assessments. Energy Rep. **6**, 173–208 (2020)
11. A.A. Elbaset, S.A.M. Abdelwahab, H.A. Ibrahim, M.A.E. Eid, *Performance Analysis of Photovoltaic Systems with Energy Storage Systems* (Springer, 2019)
12. M.F. Akorede, H. Hizam, E. Pouresmaeil, Distributed energy resources and benefits to the environment. Renew. Sust. Energ. Rev. **14**, 724–734 (2010)
13. A.R. Gupta, A. Kumar, Deployment of distributed generation with D-FACTS in distribution system: A comprehensive analytical review. IETE J. Res., 1–18 (2019)
14. S.A. Taher, S.A. Afsari, Optimal location and sizing of DSTATCOM in distribution systems by immune algorithm. Int. J. Electr. Power Energy Syst. **60**, 34–44 (2014)



A. M. A. El-Hameid et al., *Enhancement of Grid-Connected Photovoltaic Systems Using Artificial Intelligence*, https://doi.org/10.1007/978-3-031-29692-5

15. N. Salman, A. Mohamed, H. Shareef, Reliability improvement in distribution systems by optimal placement of DSTATCOM using binary gravitational search algorithm. Prz. Elektrotech. **88**, 295–299 (2012)
16. M. Sedighizadeh, A. Eisapour-Moarref, The imperialist competitive algorithm for optimal multi-objective location and sizing of DSTATCOM in distribution systems considering loads uncertainty. INAE Lett. **2**, 83–95 (2017)
17. D. Kaliaperumal Rukmani, Y. Thangaraj, U. Subramaniam, S. Ramachandran, R. Madurai Elavarasan, N. Das, L. Baringo, M.I.A. Rasheed, A new approach to optimal location and sizing of DSTATCOM in radial distribution networks using bio-inspired cuckoo search algorithm. Energies **13**, 4615 (2020)
18. P. Balamurugan, T. Yuvaraj, P. Muthukannan, Optimal allocation of DSTATCOM in distribution network using whale optimization algorithm. Eng. Technol. Appl. Sci. Res. **8**, 3445–3449 (2018)
19. A. Selim, S. Kamel, F. Jurado, Optimal allocation of distribution static compensators using a developed multi-objective sine cosine approach. Comput. Electr. Eng. **85**, 106671 (2020)
20. M. Mohammadi, M. Abasi, A.M. Rozbahani, Fuzzy-GA based algorithm for optimal placement and sizing of distribution static compensator (DSTATCOM) for loss reduction of distribution network considering reconfiguration. J. Cent. South Univ. **24**, 245–258 (2017)
21. K. Zou, A.P. Agalgaonkar, K.M. Muttaqi, S. Perera, Distribution system planning with incorporating DG reactive capability and system uncertainties. IEEE Trans. Sustain. Energy **3**, 112–123 (2011)
22. Q. Gong, J. Lei, J. Ye, Optimal siting and sizing of distributed generators in distribution systems considering cost of operation risk. Energies **9**, 61 (2016)
23. M. Ebeed, A. Alhejji, S. Kamel, F. Jurado, Solving the optimal reactive power dispatch using marine predators algorithm considering the uncertainties in load and wind-solar generation systems. Energies **13**, 4316 (2020)
24. M. Ebeed, A. Ali, M.I. Mosaad, S. Kamel, An improved lightning attachment procedure optimizer for optimal reactive power dispatch with uncertainty in renewable energy resources. IEEE Access **8**, 168721–168731 (2020)
25. M. Khajevand, A. Fakharian, M. Sedighizadeh, Stochastic joint optimal distributed generation scheduling and distribution feeder reconfiguration of microgrids considering uncertainties modeled by copula-based method. Iran. J. Electr. Electron. Eng. **16**, 371–392 (2020)
26. L. Luo, S.S. Abdulkareem, A. Rezvani, M.R. Miveh, S. Samad, N. Aljojo, M. Pazhoohesh, Optimal scheduling of a renewable based microgrid considering photovoltaic system and battery energy storage under uncertainty. J. Energy Storage **28**, 101306 (2020)
27. F.S. Abu-Mouti, M. El-Hawary, Optimal distributed generation allocation and sizing in distribution systems via artificial bee colony algorithm. IEEE Trans. Power Deliv. **26**, 2090–2101 (2011)
28. A. El-Fergany, Optimal allocation of multi-type distributed generators using backtracking search optimization algorithm. Int. J. Electr. Power Energy Syst. **64**, 1197–1205 (2015)
29. E. Ali, S.A. Elazim, A. Abdelaziz, Ant lion optimization algorithm for renewable distributed generations. Energy **116**, 445–458 (2016)
30. K. Ravi, M. Rajaram, Optimal location of FACTS devices using improved particle swarm optimization. Int. J. Electr. Power Energy Syst. **49**, 333–338 (2013)
31. O.P. Mahela, A.G. Shaik, A review of distribution static compensator. Renew. Sust. Energ. Rev. **50**, 531–546 (2015)
32. M. Mohammadi, M. Montazeri, S. Abasi, Bacterial graphical user interface oriented by particle swarm optimization strategy for optimization of multiple type DFACTS for power quality enhancement in distribution system. J. Cent. South Univ. **24**, 569–588 (2017)
33. N. Kanwar, N. Gupta, K. Niazi, A. Swarnkar, Improved cat swarm optimization for simultaneous allocation of DSTATCOM and DGs in distribution systems. J. Renew. Energy **2015** (2015)

34. T. Yuvaraj, K. Ravi, Multi-objective simultaneous DG and DSTATCOM allocation in radial distribution networks using cuckoo searching algorithm. Alex. Eng. J. **57**, 2729–2742 (2018)
35. S. Devi, M. Geethanjali, Optimal location and sizing determination of Distributed Generation and DSTATCOM using Particle Swarm Optimization algorithm. Int. J. Electr. Power Energy Syst. **62**, 562–570 (2014)
36. K. Devabalaji, K. Ravi, Optimal size and siting of multiple DG and DSTATCOM in radial distribution system using Bacterial Foraging Optimization Algorithm. Ain Shams Eng. J. **7**, 959–971 (2016)
37. T. Ackermann, G. Andersson, L. Söder, Distributed generation: A definition. Electr. Power Syst. Res. **57**, 195–204 (2001)
38. G. Allan, I. Eromenko, M. Gilmartin, I. Kockar, P. McGregor, The economics of distributed energy generation: A literature review. Renew. Sust. Energ. Rev. **42**, 543–556 (2015)
39. K. Angelopoulos, *Book Title: Integration of Distributed Generation in Low Voltage Networks: Power Quality and Economics* (ed: University of Strathclyde in Glasgow, 2004)
40. M.F. Akorede, H. Hizam, I. Aris, M.-Z.A.A. Kadir, Re-emergence of distributed generation in electric power systems: Incentives, values and issues. Energy Environ. **21**, 75–92 (2010)
41. F. Gonzalez-Longatt, C. Fortoul, Review of the distributed generation concept: Attempt of unification, in *International Conference on Renewable Energies and Power Quality (ICREPQ 05), España*, (2005), pp. 16–18
42. P. Chiradeja, Benefit of distributed generation: A line loss reduction analysis, in *2005 IEEE/PES Transmission & Distribution Conference & Exposition: Asia and Pacific*, (2005), pp. 1–5
43. B.A. Kumar, A.A. Rao, Review on resolution of issues in reactive power and voltage STATCOM based distributed generation. J. Recent Trends Power Syst. **3** (2020)
44. Y. Gao, X. Meng, W. Gao, E. Long, A review of technologies and evaluation softwares for distributed energy source system. Procedia Soc. Behav. Sci. **216**, 398–408 (2016)
45. G. Magdy, E.A. Mohamed, G. Shabib, A.A. Elbaset, Y. Mitani, SMES based a new PID controller for frequency stability of a real hybrid power system considering high wind power penetration. IET Renew. Power Gener. **12**, 1304–1313 (2018)
46. J. Martin, Distributed vs. centralized electricity generation: Are we witnessing a change of paradigm, in *An Introduction to Distributed Generation*, (2009)
47. N.E. Musoni, *Analysis of the Effect of Renewable Generation on the Power Quality of the Grid, Modelling and Analysis of Harmonic and Voltage Distortion* (Cape Peninsula University of Technology, 2018)
48. Y.-C. Jeong, E.-B. Lee, D. Alleman, Reducing voltage volatility with step voltage regulators: A life-cycle cost analysis of Korean solar photovoltaic distributed generation. Energies **12**, 652 (2019)
49. A. Ghosh, G. Ledwich, *Power Quality Enhancement Using Custom Power Devices* (Springer, 2012)
50. J.F. Hauer, J.E. Dagle, *Consortium for Electric Reliability Technology Solutions Grid of the Future White Paper on Review of Recent Reliability Issues and Systems Events* (Pacific Northwest National Lab. (PNNL), Richland, WA, 1999)
51. A. Zangeneh, S. Jadid, A. Rahimi-Kian, A hierarchical decision making model for the prioritization of distributed generation technologies: A case study for Iran. Energy Policy **37**, 5752–5763 (2009)
52. G. Ault, J. McDonald, G. Burt, Strategic analysis framework for evaluating distributed generation and utility strategies. IEE Proc. Gener. Transm. Distrib. **150**, 475–481 (2003)
53. M. Zhang, L. Dai, Carbon nanomaterials as metal-free catalysts in next generation fuel cells. Nano Energy **1**, 514–517 (2012)
54. T. Khatib, W. Elmenreich, *Modeling of Photovoltaic Systems Using Matlab: Simplified Green Codes* (Wiley, 2016)
55. S.-K. Kim, J.-H. Jeon, C.-H. Cho, E.-S. Kim, J.-B. Ahn, Modeling and simulation of a grid-connected PV generation system for electromagnetic transient analysis. Sol. Energy **83**, 664–678 (2009)

56. M. Abd El Sattar, A.M. Abd El Hamed, A.A. Elbaset, M. Ebeed, Application of enhanced sine cosine algorithm for optimal allocation of PV-DG and DSTATCOM in distribution systems. Int. J. Eng. Manag. Res. **10**(3), 594–615 (2021)
57. R. Sirjani, A.R. Jordehi, Optimal placement and sizing of distribution static compensator (D-STATCOM) in electric distribution networks: A review. Renew. Sust. Energ. Rev. **77**, 688–694 (2017)
58. S. Elango, E.C. Sekaran, Mitigation of Voltage Sag by Using Distribution Static Compensator (D-STATCOM), in *2011 International Conference on Process Automation, Control and Computing*, (2011), pp. 1–6
59. S.A. Taher, A.R. Afsari, Optimal allocation and sizing of dstatcom by immune algorithm in distribution networks including distributed generation. Soft Comput. J. **2**, 2–15 (2013)
60. A.R. Jordehi, Particle swarm optimisation (PSO) for allocation of FACTS devices in electric transmission systems: A review. Renew. Sust. Energ. Rev. **52**, 1260–1267 (2015)
61. A.R. Gupta, A. Kumar, Reactive power deployment and cost benefit analysis in DNO operated distribution electricity markets with D-STATCOM. Front. Energy **13**, 86–98 (2019)
62. M. Asensio, G. Muñoz-Delgado, J. Contreras, Bi-level approach to distribution network and renewable energy expansion planning considering demand response. IEEE Trans. Power Syst. **32**, 4298–4309 (2017)
63. P. Prakash, D.K. Khatod, Optimal sizing and siting techniques for distributed generation in distribution systems: A review. Renew. Sust. Energ. Rev. **57**, 111–130 (2016)
64. B. Sultana, M. Mustafa, U. Sultana, A.R. Bhatti, Review on reliability improvement and power loss reduction in distribution system via network reconfiguration. Renew. Sust. Energ. Rev. **66**, 297–310 (2016)
65. A.R. Jordehi, Particle swarm optimisation with opposition learning-based strategy: An efficient optimisation algorithm for day-ahead scheduling and reconfiguration in active distribution systems. Soft. Comput. **24**, 18573–18590 (2020)
66. A.R. Jordehi, Allocation of distributed generation units in electric power systems: A review. Renew. Sust. Energ. Rev. **56**, 893–905 (2016)
67. A.R. Jordehi, Brainstorm optimisation algorithm (BSOA): An efficient algorithm for finding optimal location and setting of FACTS devices in electric power systems. Int. J. Electr. Power Energy Syst. **69**, 48–57 (2015)
68. H.-J. Choi, J.-H. Jung, Enhanced power line communication strategy for DC microgrids using switching frequency modulation of power converters. IEEE Trans. Power Electron. **32**, 4140–4144 (2017)
69. F. Dastgeer, H.E. Gelani, A comparative analysis of system efficiency for AC and DC residential power distribution paradigms. Energ. Buildings **138**, 648–654 (2017)
70. Á. Molina-García, R.A. Mastromauro, T. García-Sánchez, S. Pugliese, M. Liserre, S. Stasi, Reactive power flow control for PV inverters voltage support in LV distribution networks. IEEE Trans. Smart Grid **8**, 447–456 (2016)
71. R. Majumder, Reactive power compensation in single-phase operation of microgrid. IEEE Trans. Ind. Electron. **60**, 1403–1416 (2012)
72. A. Elnady, M.M. Salama, Unified approach for mitigating voltage sag and voltage flicker using the DSTATCOM. IEEE Trans. Power Deliv. **20**, 992–1000 (2005)
73. Z. Shuai, A. Luo, Z.J. Shen, W. Zhu, Z. Lv, C. Wu, A dynamic hybrid var compensator and a two-level collaborative optimization compensation method. IEEE Trans. Power Electron. **24**, 2091–2100 (2009)
74. R. Yan, B. Marais, T.K. Saha, Impacts of residential photovoltaic power fluctuation on on-load tap changer operation and a solution using DSTATCOM. Electr. Power Syst. Res. **111**, 185–193 (2014)
75. M. Hosseini, H.A. Shayanfar, Regular paper modeling of series and shunt distribution FACTS devices in distribution systems load flow. J. Electr. Syst. **4**, 1–12 (2008)
76. M. Ebeed, S. Kamel, S.H.A. Aleem, A.Y. Abdelaziz, Optimal allocation of compensators, in *Electric Distribution Network Planning*, (Springer, 2018), pp. 321–353

77. T.L. Skvarenina, *The Power Electronics Handbook* (CRC Press, 2018)
78. A.A. Elbaset, H. Ali, M. Abd-El Sattar, Novel seven-parameter model for photovoltaic modules. Sol. Energy Mater. Sol. Cells **130**, 442–455 (2014)
79. A.A. Elbaset, H. Ali, M. Abd El Sattar, New seven parameters model for amorphous silicon and thin film PV modules based on solar irradiance. Sol. Energy **138**, 26–35 (2016)
80. T. Salmi, M. Bouzguenda, A. Gastli, A. Masmoudi, Matlab/simulink based modeling of photovoltaic cell. Int. J. Renew. Energy Res. **2**, 213–218 (2012)
81. A.A. Elbaset, H. Ali, M. Abd-El Sattar, M. Khaled, Implementation of a modified perturb and observe maximum power point tracking algorithm for photovoltaic system using an embedded microcontroller. IET Renew. Power Gener. **10**, 551–560 (2016)
82. C.-C. Chu, C.-L. Chen, Robust maximum power point tracking method for photovoltaic cells: A sliding mode control approach. Sol. Energy **83**, 1370–1378 (2009)
83. A.A. Elbaset, M. Hassan, H. Ali, Performance analysis of grid-connected PV system, in *2016 Eighteenth International Middle East Power Systems Conference (MEPCON)*, (2016), pp. 675–682
84. W.B. Walker, E.M. Smith, T. Jan, L. Zwiebel, A functional role for Anopheles gambiae Arrestin1 in olfactory signal transduction. J. Insect Physiol. **54**, 680–690 (2008)
85. B.S. Bernanke, Report of the Editor: American economic review. Am. Econ. Rev. **94**, 501–513 (2004)
86. S. Liu, L. Wang, J. Tian, Y. Luo, X. Zhang, X. Sun, Aniline as a dispersing and stabilizing agent for reduced graphene oxide and its subsequent decoration with Ag nanoparticles for enzymeless hydrogen peroxide detection. J. Colloid Interface Sci. **363**, 615–619 (2011)
87. M.A. Akbari, J. Aghaei, M. Barani, M. Savaghebi, M. Shafie-Khah, J.M. Guerrero, J.P. Catalao, New metrics for evaluating technical benefits and risks of DGs increasing penetration. IEEE Trans. Smart Grid **8**, 2890–2902 (2017)
88. R.H. Zubo, G. Mokryani, R.J.A.E. Abd-Alhameed, Optimal operation of distribution networks with high penetration of wind and solar power within a joint active and reactive distribution market environment. Appl. Energy **220**, 713–722 (2018)
89. M. Eremia, C.-C. Liu, A.-A. Edris, *Advanced Solutions in Power Systems: HVDC, FACTS, and Artificial Intelligence* (Wiley, 2016)
90. A. Edris, Proposed terms and definitions for flexible AC transmission system (FACTS). IEEE Trans. Power Deliv. **12** (1997)
91. C. Liu, J. Masri, V. Perez, C. Maya, J. Zhao, Growth performance and nutrient composition of mealworms (Tenebrio molitor) fed on fresh plant materials-supplemented diets. Foods **9**, 151 (2020)
92. S. Moghaddas-Tafreshi, E. Mashhour, Distributed generation modeling for power flow studies and a three-phase unbalanced power flow solution for radial distribution systems considering distributed generation. Electr. Power Syst. Res. **79**, 680–686 (2009)
93. M. Emmanuel, R. Rayudu, Evolution of dispatchable photovoltaic system integration with the electric power network for smart grid applications: A review. Renew. Sust. Energ. Rev. **67**, 207–224 (2017)
94. R.M. Neve, K. Chin, J. Fridlyand, J. Yeh, F.L. Baehner, T. Fevr, L. Clark, N. Bayani, J.-P. Coppe, F. Tong, A collection of breast cancer cell lines for the study of functionally distinct cancer subtypes. Cancer Cell **10**, 515–527 (2006)
95. J.M. Solanki, S. Khushalani, N.N. Schulz, A multi-agent solution to distribution systems restoration. IEEE Trans. Power Syst. **22**, 1026–1034 (2007)
96. H. Johal, D. Divan, Design considerations for series-connected distributed FACTS converters. IEEE Trans. Ind. Appl. **43**, 1609–1618 (2007)
97. L. Gyugyi, Dynamic compensation of AC transmission lines by solid-state synchronous voltage sources. IEEE Trans. Power Deliv. **9**, 904–911 (1994)
98. J. Savier, D. Das, Impact of network reconfiguration on loss allocation of radial distribution systems. IEEE Trans. Power Deliv. **22**, 2473–2480 (2007)

99. J.P. Lopes, N. Hatziargyriou, J. Mutale, P. Djapic, N. Jenkins, Integrating distributed generation into electric power systems: A review of drivers, challenges and opportunities. Electr. Power Syst. Res. **77**, 1189–1203 (2007)
100. S.A. Papathanassiou, A technical evaluation framework for the connection of DG to the distribution network. Electr. Power Syst. Res. **77**, 24–34 (2007)
101. D. Divan, W. Brumsickle, R. Schneider, B. Kranz, R. Gascoigne, D. Bradshaw, M. Ingram, I. Grant, A distributed static series compensator system for realizing active power flow control on existing power lines, in *IEEE PES Power Systems Conference and Exposition, 2004*, (2004), pp. 654–661
102. M.H. Golshan, S. Arefifar, Distributed generation, reactive sources and network-configuration planning for power and energy-loss reduction. IEE Proc. Gener. Transm. Distrib. **153**, 127–136 (2006)
103. N. Hadjsaid, J.-F. Canard, F. Dumas, Dispersed generation impact on distribution networks. IEEE Comput. Appl. Power **12**, 22–28 (1999)
104. A.A. Elbaset, M.S. Hassan, *Design and Power Quality Improvement of Photovoltaic Power System* (Springer, 2017)
105. K. Suslov, N. Solonina, D. Gerasimov, Assessment of an impact of power supply participants on power quality, in *2018 18th International Conference on Harmonics and Quality of Power (ICHQP)*, (2018), pp. 1–5
106. A. Delle Femine, D. Gallo, D. Giordano, C. Landi, M. Luiso, D. Signorino, Power quality assessment in railway traction supply systems. IEEE Trans. Instrum. Meas. **69**, 2355–2366 (2020)
107. W. Xu, X. Liu, Y. Liu, An investigation on the validity of power-direction method for harmonic source determination. IEEE Trans. Power Deliv. **18**, 214–219 (2003)
108. S. Khalid, B. Dwivedi, Power quality issues, problems, standards & their effects in industry with corrective means. Int. J. Adv. Eng. Technol. **1**, 1 (2011)
109. B.S. GOUD, B.L. Rao, An intelligent technique for optimal power quality enhancement (OPQE) in a HRES grid connected system: ESA technique. Int. J. Renew. Energy Res. **10**, 317–328 (2020)
110. N. Prakash, V. Balaji, M. Sudha, Power quality improvement of grid inter connected hybrid system using STATCOM. Int. J. Adv. Eng. Technol. **7**, 1225–1233 (2016)
111. S. Angalaeswari, M. Thejeswar, R.S. Poongodi, W.V. Basha, P. Sasikumar, Power quality improvement of standalone hybrid solar-wind power generation system using FACTS devices. Int. J. Adv. Res. Electr. Electron. Instrum. Eng., (An ISO 3297: 2007 Certified Organization), **3** (2014)
112. R.D. Singh, T.P. Kumar, K. Sumanth, Simulation of SRF based DSTATCOM with grid connected PV generation system using fuzzy logic controller for reactive power management. Int. J. Adv. Res. Electr. Electron. Instrum. Eng. **5**, 6493–6501 (2016)
113. D. Amoozegar, DSTATCOM modelling for voltage stability with fuzzy logic PI current controller. Int. J. Electr. Power Energy Syst. **76**, 129–135 (2016)
114. A. Moreno-Muñoz, *Power Quality: Mitigation Technologies in a Distributed Environment* (Springer, 2007)
115. C.A.G. Marques, D.D. Ferreira, L.R. Freitas, C.A. Duque, M.V. Ribeiro, Improved disturbance detection technique for power-quality analysis. IEEE Trans. Power Deliv. **26**, 1286–1287 (2011)
116. M. Premkumar, C. Kumar, R. Sowmya, Mathematical modelling of solar photovoltaic cell/panel/array based on the physical parameters from the manufacturer's datasheet. Int. J. Renew. Energy Dev. **9** (2020)
117. M. Aourir, A. Abouloifa, I. Lachkar, A. Hamdoun, F. Giri, F. Cuny, Nonlinear control of PV system connected to single phase grid through half bridge power inverter. IFAC-PapersOnLine **50**, 741–746 (2017)
118. H.-C. Chu, G.-J. Hwang, C.-C. Tsai, A knowledge engineering approach to developing mindtools for context-aware ubiquitous learning. Comput. Educ. **54**, 289–297 (2010)

119. R.K. Nema, S. Nema, G. Agnihotri, Design, development and simulation of PC-based scheme for characterisation of solar photovoltaic modules. Int. J. Power Electron. **2**, 304–325 (2010)
120. K. Prakash, N. Nadkarni, W.K. Lye, M.H. Yong, E.K. Tan, The impact of non-motor symptoms on the quality of life of Parkinson's disease patients: A longitudinal study. Eur. J. Neurol. **23**, 854–860 (2016)
121. S. Angalaeswari, O.G. Swathika, V. Ananthakrishnan, J.F. Daya, K. Jamuna, Efficient power management of grid operated microgrid using fuzzy logic controller (FLC). Energy Procedia **117**, 268–274 (2017)
122. S. Angalaeswari, K. Jamuna, Modified iterative learning controller for efficient power management of hybrid AC/DC microgrid. Int. J. Innov. Comput. Appl. **12**, 24–36 (2021)
123. B. Thangaraj, K.B. Ramachandran, S.P. Raj, Homogeneous catalytic transesterification of renewable Azadirachta indica (Neem) oil and its derivatives to biodiesel fuel via acid/alkaline esterification processes. Int. J. Renew. Energy Biofuels **11**, 1–16 (2014)
124. A. Laguna, S. Aranda, M.J. Barallobre, R. Barhoum, E. Fernández, V. Fotaki, J.M. Delabar, S. de la Luna, P. de la Villa, M.L. Arbonés, The protein kinase DYRK1A regulates caspase-9-mediated apoptosis during retina development. Dev. Cell **15**, 841–853 (2008)
125. M.C. Falvo, L. Martirano, D. Sbordone, E. Bocci, Technologies for smart grids: A brief review, in *2013 12th International Conference on Environment and Electrical Engineering*, (2013), pp. 369–375
126. S.S. Khadake, N.M. Agashe, A.P. Kinage, New control technique to improve the grid power by using PV-STATCOM, in *2018 International Conference on Information, Communication, Engineering and Technology (ICICET)*, (2018), pp. 1–5
127. T.U. Okeke, R.G. Zaher, Flexible AC transmission systems (FACTS), in *2013 International Conference on New Concepts in Smart Cities: Fostering Public and Private Alliances (SmartMILE)*, (2013), pp. 1–4
128. L.N. Popavath, P. Kaliannan, Photovoltaic-STATCOM with low voltage ride through strategy and power quality enhancement in a grid integrated wind-PV system. Electronics **7**, 51 (2018)
129. S. Alzahrani, R. Shah, N. Mithulananthan, A. Sode-Yome, Large-scale PV voltage regulation: Survey of recent practice, in *2019 IEEE PES GTD Grand International Conference and Exposition Asia (GTD Asia)*, (2019), pp. 661–666
130. R.K. Varma, E.M. Siavashi, PV-STATCOM: A new smart inverter for voltage control in distribution systems. IEEE Trans. Sustain. Energy **9**, 1681–1691 (2018)
131. J.M. Rupa, S. Ganesh, Power flow analysis for radial distribution system using backward/forward sweep method. Int. J. Electr. Comp. Electron. Commun. Eng. **8**, 1540–1544 (2014)
132. R. Heffernan, F. Mostashari, D. Das, A. Karpati, M. Kulldorff, D. Weiss, Syndromic surveillance in public health practice, New York City. Emerg. Infect. Dis. **10**, 858 (2004)
133. R.A. Jabr, Optimal power flow using an extended conic quadratic formulation. IEEE Trans. Power Syst. **23**, 1000–1008 (2008)
134. J. Tu, Z. Yin, Y. Xu, Study on the evaluation index system and evaluation method of voltage stability of distribution network with high DG penetration. Energies **11**, 79 (2018)
135. J.G. Inoue, M. Miya, K. Lam, B.-H. Tay, J.A. Danks, J. Bell, T.I. Walker, B. Venkatesh, Evolutionary origin and phylogeny of the modern holocephalans (Chondrichthyes: Chimaeriformes): A mitogenomic perspective. Mol. Biol. Evol. **27**, 2576–2586 (2010)
136. A.A. Abou El-Ela, R.A. El-Sehiemy, A.-M. Kinawy, M.T. Mouwafi, Optimal capacitor placement in distribution systems for power loss reduction and voltage profile improvement. IET Gener. Transm. Distrib. **10**, 1209–1221 (2016)
137. S. Sivanagaraju, N. Visali, V. Sankar, T. Ramana, Enhancing voltage stability of radial distribution systems by network reconfiguration. Electr. Power Compon. Syst. **33**, 539–550 (2005)
138. M. Chakravorty, D. Das, Voltage stability analysis of radial distribution networks. Int. J. Electr. Power Energy Syst. **23**, 129–135 (2001)
139. M. Aman, G. Jasmon, A. Bakar, H. Mokhlis, A new approach for optimum simultaneous multi-DG distributed generation Units placement and sizing based on maximization of system

loadability using HPSO (hybrid particle swarm optimization) algorithm. Energy **66**, 202–215 (2014)
140. A. Ali, D. Raisz, K. Mahmoud, M. Lehtonen, Optimal placement and sizing of uncertain PVs considering stochastic nature of PEVs. IEEE Trans. Sustain. Energy **11**, 1647–1656 (2019)
141. A. Ali, K. Mahmoud, D. Raisz, M. Lehtonen, Probabilistic approach for hosting high PV penetration in distribution systems via optimal oversized inverter with watt-var functions. IEEE Syst. J. **15**, 684–693 (2020)
142. L. Luo, W. Gu, X.-P. Zhang, G. Cao, W. Wang, G. Zhu, D. You, Z. Wu, Optimal siting and sizing of distributed generation in distribution systems with PV solar farm utilized as STATCOM (PV-STATCOM). Appl. Energy **210**, 1092–1100 (2018)
143. Z. Luo, S. Yang, N. Xie, W. Xie, J. Liu, Y.S. Agbodjan, Z. Liu, Multi-objective capacity optimization of a distributed energy system considering economy, environment and energy. Energy Convers. Manag. **200**, 112081 (2019)
144. A. Faramarzi, M. Heidarinejad, S. Mirjalili, A.H. Gandomi, Marine predators algorithm: A nature-inspired metaheuristic. Expert Syst. Appl. **152**, 113377 (2020)
145. A. Faramarzi, M. Heidarinejad, B. Stephens, S. Mirjalili, Equilibrium optimizer: A novel optimization algorithm. Knowl.-Based Syst. **191**, 105190 (2020)
146. S. Mirjalili, The ant lion optimizer. Adv. Eng. Softw. **83**, 80–98 (2015)
147. P.R. Hof, E. Van der Gucht, Structure of the cerebral cortex of the humpback whale, Megaptera novaeangliae (Cetacea, Mysticeti, Balaenopteridae). Anat. Rec. Adv. Integr. Anat. Evol. Biol. **290**, 1–31 (2007)
148. W.A. Watkins, W.E. Schevill, Aerial observation of feeding behavior in four baleen whales: Eubalaena glacialis, Balaenoptera borealis, Megaptera novaeangliae, and Balaenoptera physalus. J. Mammal. **60**, 155–163 (1979)
149. J.A. Goldbogen, A.S. Friedlaender, J. Calambokidis, M.F. Mckenna, M. Simon, D.P. Nowacek, Integrative approaches to the study of baleen whale diving behavior, feeding performance, and foraging ecology. Bioscience **63**, 90–100 (2013)
150. R. Lin, Y. Xia, A new eigensolution of structures via dynamic condensation. J. Sound Vib. **266**, 93–106 (2003)
151. S. Li, H. Chen, M. Wang, A.A. Heidari, S. Mirjalili, Slime mould algorithm: A new method for stochastic optimization. Futur. Gener. Comput. Syst. **111**, 300–323 (2020)
152. N. Sahoo, K. Prasad, A fuzzy genetic approach for network reconfiguration to enhance voltage stability in radial distribution systems. Energy Convers. Manag. **47**, 3288–3306 (2006)
153. N. Acharya, P. Mahat, N. Mithulananthan, An analytical approach for DG allocation in primary distribution network. Int. J. Electr. Power Energy Syst. **28**, 669–678 (2006)
154. K. Mahmoud, N. Yorino, A. Ahmed, Optimal distributed generation allocation in distribution systems for loss minimization. IEEE Trans. Power Syst. **31**, 960–969 (2015)
155. J.A.M. García, A.J.G. Mena, Optimal distributed generation location and size using a modified teaching–learning based optimization algorithm. Int. J. Electr. Power Energy Syst. **50**, 65–75 (2013)
156. W. Tan, M. Hassan, M. Majid, H. Rahman, Allocation and sizing of DG using cuckoo search algorithm, in *2012 IEEE International Conference on Power and Energy (PECon)*, (2012), pp. 133–138
157. S. Kansal, V. Kumar, B. Tyagi, Hybrid approach for optimal placement of multiple DGs of multiple types in distribution networks. Int. J. Electr. Power Energy Syst. **75**, 226–235 (2016)
158. I. Pisica, C. Bulac, M. Eremia, Optimal distributed generation location and sizing using genetic algorithms, in *2009 15th International Conference on Intelligent System Applications to Power Systems*, (2009), pp. 1–6
159. S. Kaur, G. Kumbhar, J. Sharma, A MINLP technique for optimal placement of multiple DG units in distribution systems. Int. J. Electr. Power Energy Syst. **63**, 609–617 (2014)
160. T. Shukla, S. Singh, V. Srinivasarao, K. Naik, Optimal sizing of distributed generation placed on radial distribution systems. Electr. Power Compon. Syst. **38**, 260–274 (2010)

161. S. Kamel, A. Amin, A. Selim, M.H. Ahmed, Optimal placement of DG and capacitor in radial distribution systems considering load variation, in *2019 International Conference on Computer, Control, Electrical, and Electronics Engineering (ICCCEEE)*, (2019), pp. 1–6
162. S.R. Gampa, D. Das, Optimum placement and sizing of DGs considering average hourly variations of load. Int. J. Electr. Power Energy Syst. **66**, 25–40 (2015)
163. A.R. Gupta, A. Kumar, Performance analysis of radial distribution systems with UPQC and D-STATCOM. J. Inst. Eng. (India): B **98**, 415–422 (2017)
164. S. Sultana, P.K. Roy, Optimal capacitor placement in radial distribution systems using teaching learning based optimization. Int. J. Electr. Power Energy Syst. **54**, 387–398 (2014)
165. D.F. Pires, C.H. Antunes, A.G. Martins, NSGA-II with local search for a multi-objective reactive power compensation problem. Int. J. Electr. Power Energy Syst. **43**, 313–324 (2012)
166. A. El-Fergany, Study impact of various load models on DG placement and sizing using backtracking search algorithm. Appl. Soft Comput. **30**, 803–811 (2015)
167. D. Thukaram, H.W. Banda, J. Jerome, A robust three phase power flow algorithm for radial distribution systems. Electr. Power Syst. Res. **50**, 227–236 (1999)
168. G. Chang, S. Chu, H. Wang, An improved backward/forward sweep load flow algorithm for radial distribution systems. IEEE Trans. Power Syst. **22**, 882–884 (2007)
169. A.A. Hassan, F.H. Fahmy, A.E.-S.A. Nafeh, M.A. Abu-elmagd, Genetic single objective optimisation for sizing and allocation of renewable DG systems. Int. J. Sustain. Energy **36**, 545–562 (2017)
170. H. Manafi, N. Ghadimi, M. Ojaroudi, P. Farhadi, Optimal placement of distributed generations in radial distribution systems using various PSO and DE algorithms. Elektron. ir Elektrotech. **19**, 53–57 (2013)
171. D.Q. Hung, N. Mithulananthan, Multiple distributed generator placement in primary distribution networks for loss reduction. IEEE Trans. Ind. Electron. **60**, 1700–1708 (2011)
172. M. Kefayat, A.L. Ara, S.N. Niaki, A hybrid of ant colony optimization and artificial bee colony algorithm for probabilistic optimal placement and sizing of distributed energy resources. Energy Convers. Manag. **92**, 149–161 (2015)
173. D. Zhang, Z. Fu, L. Zhang, An improved TS algorithm for loss-minimum reconfiguration in large-scale distribution systems. Electr. Power Syst. Res. **77**, 685–694 (2007)

# Index


**A**

Annual cost, 2, 3, 44, 47, 132, 134, 165, 166
Ant lion optimizer (ALO), 3, 55, 56, 60, 62–64, 86–90, 101–118, 160, 161, 165, 193, 207, 208

**C**

Capital costs, 9–10, 15, 21, 48, 50

**D**

Distributed energy, 30, 142, 144
Distributed energy resources (DERs), 3, 6, 29, 45, 51, 54, 73–99, 104–130, 139–159
Distributed flexible AC transmission system (D-FACTS), 2, 3, 5, 6, 28–29, 37, 39
Distributed generation, 1, 5, 7–17, 36, 37
Distributed network, 39
Distributed static compensator (D-STATCOM), 2, 3, 5–7, 16–21, 31, 37, 39–71, 76, 77, 79, 80, 82–84, 101–104, 109–112, 115, 116, 119, 121–134, 136–141, 151, 157, 158, 160, 161, 165, 166, 179, 184, 191, 200, 207

**E**

Enhanced sine cosine algorithm (ESCA), 3, 84–87, 142–159
Equilibrium optimizer (EO), 3, 76–80, 116–129, 160, 161, 165

**F**

Flexible alternating current transmission system (FACTS) devices, 17, 26–29, 31
Forward/backward, 39–44, 54, 55, 142, 165

**I**

IEEE-30 bus, 55
IEEE-33 bus, 3, 4
IEEE-69 bus, 3, 4, 107–114, 151, 165, 166
IEEE-94 bus, 4, 165, 166
IEEE-118 bus, 3, 4, 104, 114–118, 160–161, 165, 166

**L**

Lightning attachment procedure optimization (LAPO), 3, 80–84, 116–129, 160, 161, 165

**M**

Marine predators algorithm (MPA), 3, 73, 76, 77, 129–130, 132–134, 136, 137, 160, 161, 165
Modified ant lion optimizer (MALO), 3, 90–91, 104–118, 160, 161

**P**

Penetration level, 22, 52, 121–126, 128, 129, 166




A. M. A. El-Hameid et al., *Enhancement of Grid-Connected Photovoltaic Systems Using Artificial Intelligence*, https://doi.org/10.1007/978-3-031-29692-5

Photovoltaic (PV), 1, 2, 5, 6, 9, 10, 15, 17, 18, 21–29, 31, 37, 44, 47, 48, 50–55, 60, 62, 79, 80, 82–84, 101–105, 107–111, 113–119, 121–129, 131, 132, 134, 137–141, 143–146, 149, 151–153, 155–158, 160, 161, 189–191, 205–207
Photovoltaic distributed generation (PV-DG), 1–3, 6, 16, 25, 27–28, 37, 39, 44, 48, 76, 77, 104, 109–112, 116, 129–134, 136–139, 143, 151–159, 165, 166
Power losses, 2, 3, 5, 7, 16–18, 39, 44, 45, 55, 56, 60, 64, 101, 105, 109, 110, 114, 119, 121, 123, 124, 127, 130–132, 134, 139–141, 143, 144, 151, 155, 156, 165
Power quality, 2, 3, 5–7, 12, 16–18, 30–37, 39
Power quality issues, 2, 3, 5–37

**R**
Radial distribution network (RDN), 5–7, 43, 165

**S**
Sensitivity analysis, 54–71, 79, 80, 82, 83, 116, 119, 165
Slime mould algorithm (SMA), 3, 95–99, 139–141, 160, 161, 165

**U**
Uncertainty, 3, 6, 53, 107–118, 131–139, 160, 166
Uncertainty modeling, 53–54

**V**
Voltage deviation, 3, 5, 7, 17, 44, 46, 55, 56, 60, 61, 64, 132, 134, 151, 160, 166
Voltage profile, 6, 7, 16, 17, 39, 56, 61, 64, 101, 103, 107, 114–117, 134, 135, 141, 143, 144, 149–151, 159, 165
Voltage stability index, 3, 45, 46, 55, 60, 103, 104, 132, 149, 159, 166, 211

**W**
Whale optimization algorithm (WOA), 3, 91–96, 109, 112–114, 116–129, 160, 161, 165

MIX
Papier aus verantwortungsvollen Quellen
Paper from responsible sources
FSC® C105338
FSC
www.fsc.org

If you have any concerns about our products,
you can contact us on
**ProductSafety@springernature.com**

In case Publisher is established outside the EU,
the EU authorized representative is:
**Springer Nature Customer Service Center GmbH**
**Europaplatz 3, 69115 Heidelberg, Germany**

Printed by Libri Plureos GmbH
in Hamburg, Germany